Stable Isotopes in Nutrition

A C S S Y M P O S I U M S E R I E S **258**

Stable Isotopes in Nutrition

Judith R. Turnlund, EDITOR
Phyllis E. Johnson, EDITOR
U.S. Department of Agriculture

Based on a symposium sponsored by
the Division of Agricultural and Food Chemistry
at the 186th Meeting
of the American Chemical Society,
Washington, D.C.,
August 28–September 2, 1983

American Chemical Society, Washington, D.C. 1984

Library of Congress Cataloging in Publication Data

Stable isotopes in nutrition.
(ACS symposium series, ISSN 0097-6156; 258)
Includes bibliographies and indexes.

1. Stable isotope tracers—Congresses. 2. Nutrition—Congresses. 3. Metabolism—Research—Methodology—Congresses.

I. Turnlund, Judith R. II. Johnson, Phyllis E. III. American Chemical Society. Meeting (186th: 1983: Washington, D.C.) IV. Series. [DNLM: 1. Isotopes—congresses. 2. Metabolism—congresses. 3. Nutrition—congresses. QU 120 S775 1983]

QP143.5.S73S73 1984 612'.3'028 84-12430
ISBN 0-8412-0855-7

PRINTED IN THE UNITED STATES OF AMERICA

ACS Symposium Series

M. Joan Comstock, *Series Editor*

FOREWORD

The ACS SYMPOSIUM SERIES was founded in 1974 to provide a medium for publishing symposia quickly in book form. The format of the Series parallels that of the continuing ADVANCES IN CHEMISTRY SERIES except that in order to save time the papers are not typeset but are reproduced as they are submitted by the authors in camera-ready form. Papers are reviewed under the supervision of the Editors with the assistance of the Series Advisory Board and are selected to maintain the integrity of the symposia; however, verbatim reproductions of previously published papers are not accepted. Both reviews and reports of research are acceptable since symposia may embrace both types of presentation.

CONTENTS

PREFACE

STABLE ISOTOPES ARE NEW TOOLS FOR NUTRITIONISTS that enable researchers to label nutrients for human experiments without exposing study participants to radioactivity. In addition some elements have no suitable radioisotopes, and stable isotopes provide the only means of conducting experiments with labels. This field of research relies heavily on expertise in sophisticated chemical techniques and analytical instrumentation, and nutritionists have entered the field only recently. The field is in its infancy, but is now expanding rapidly; there is widespread interest in use of stable isotopes to study nutrient bioavailability, metabolism, and utilization. Many approaches and analytical methods are being developed for use of stable isotopes in nutrition research.

This book focuses on the chemistry and instrumentation employed in the analysis of isotopically enriched samples as well as on nutritional aspects of stable isotope research. Emphasis is placed on the stable mineral isotope area, which is new to nutrition research. Because of the novelty of the use of stable mineral isotopes in nutrition, there is yet no consensus about the best analytical methods to use for measuring isotopic enrichment. Each laboratory using stable isotopes has developed its own methods independently. Differences in sample matrices or in specific nutrients analyzed have also contributed to differences in methods. Several chapters on analysis of isotopically enriched organic nutrients and their metabolites (^{13}C and ^{15}N labels) complement chapters on mineral isotope analysis.

JUDITH R. TURNLUND
U.S. Department of Agriculture
Berkeley, California

PHYLLIS E. JOHNSON
U.S. Department of Agriculture
Grand Forks, North Dakota

March 1984

1

Stable Isotope Measurements with Thermal and Resonance Ionization Mass Spectrometry

L. J. MOORE

Center for Analytical Chemistry, National Measurement Laboratory, National Bureau of Standards, Washington, DC 20234

Thermal ionization mass spectrometry has been used extensively in the geological, nuclear and analytical sciences for stable isotope measurements. A new technique, resonance ionization mass spectrometry, offers a comprehensive approach to sensitive and selective elemental and isotopic analysis. Recent developments in thermal and resonance ionization mass spectrometry are reviewed, and specific applications of the technology to zinc and calcium metabolism studies and to trace element analysis of foodstuffs are summarized. The application potential and importance of the mass spectrometry technology to nutrition research are indicated.

Thermal ionization mass spectrometry (TIMS) has been employed extensively in the geological, nuclear and analytical chemistry communities during the last few decades to measure isotopes. A new technique, resonance ionization mass spectrometry (RIMS) has begun to emerge within the last two or three years and appears to offer a very comprehensive approach to sensitive and selective stable isotope measurements. Applications of TIMS to nutrition and bioavailability studies have begun relatively recently; in a sense, the use of stable isotopes in nutrition and bioavailability studies is poised now where it was many years ago in the other communities. Since the use of inorganic stable isotope tracers in nutrition and bioavailability studies is relatively recent, it is worthwhile to illustrate the success of mass spectrometry applications in the geological, nuclear and analytical sciences and to assess by analogy the potential impact available to research scientists in nutrition. The same chemical separations and measurement technology that have been developed for these other scientific areas can, in many cases, be transferred almost directly to stable isotope measurements in biological systems. The purposes of this paper are to summarize progress and developments in TIMS, to describe recent advances in RIMS, and to suggest how these techniques might be expected to impact nutrition research and bioavailability studies.

Since stable isotopes were first observed in nature with early mass spectroscopes (1) stable isotope mass spectrometry has grown into a mature discipline that embraces a diverse spectrum of capabilities. In the geological sciences, high precision isotope ratio measurements have resolved small quantities of stable decay products, such as ^{87}Sr and ^{143}Nd, that are derived from their long-lived radiogenic parents, ^{87}Rb and ^{147}Sm, respectively. Stable isotope decay products have been used to determine geochronological ages and to trace the mixing patterns in inter-ocean water flows. TIMS has been used to accurately measure elemental concentrations in geological samples using stable isotope dilution techniques. These applications have proven highly successful in studies of terrestrial geological processes and studies of the origin of lunar samples, the solar system, meteorites and related extraterrestrial materials. Measurements of radiogenic lead isotopes have been used to determine the provenance of archaeological artifacts and to identify sources of lead pollution. TIMS is an indispensable tool in the nuclear industry to measure isotopic nuclear burn-up products, half-lives of radioactive isotopes, and concentrations of low level nuclear contaminants in biological materials.

At the National Bureau of Standards (NBS), a long term program in highly accurate and precise stable isotope measurements has provided the measurement basis to support research and development in the geological community and nuclear industry. This technology transfer has occurred in the form of state-of-the-art instrumentation, isotopic Standard Reference Materials(SRM's), and methodology development for chemical separations and isotopic analysis. In support of other NBS programs, mass spectrometry has been used to accurately measure trace element concentrations in a wide variety of biological and environmental SRM's. The accuracy of mass spectrometry has also been applied in a separate long term program to determine the atomic weights of the elements at levels of accuracy not achievable with other measurement approaches. During the evolution of the TIMS technique, the sources of systematic bias have been carefully investigated and minimized. The near-absence of systematic measurement biases in TIMS has resulted, in recent years, in its application as a definitive measurement tool to determine concentrations of electrolytes in biological fluids using isotope dilution techniques (2-4). These definitive values were then used to calibrate other more cost-effective analytical methodologies used in clinical laboratories. Presently, a Human Serum SRM (#909) is being certified for selected constituents using TIMS and RIMS with isotope dilution technology (5).

The use of stable isotopes and TIMS in the medical and nutrition areas has begun only recently, even though the scientific potential is very great. There are several reasons for the lag in usage of inorganic (mineral) stable isotopes by the medical and nutrition community: interdisciplinary awareness; relative technical complexity; instrumentation cost; and measurement time requirements. Education of the practitioners in nutrition and mass spectrometry to the needs and capabilities of their interdisciplinary counterparts has been somewhat neglected. The complexity requires a lengthy and dedicated effort in the development of trained personnel and technical expertise. Furthermore, the cost of highly precise state-of-the-art mass spectrometry required to resolve low level tracer enrichments is beyond the budget of many laboratories. In recent years, the number of nutrition, metabolism and bioavailability studies with inorganic stable isotopes has risen dramatically, and a number of the measurements have been accomplished with neutron activation analysis (NAA) (6-8) and TIMS (8-10). A trend away from the use of radioactive tracers for such studies in newborns, children and pregnant women has heightened the interest in using stable isotopes. The use of NAA and TIMS for such studies has occurred predominantly in large institutions where existing technical capabilities, normally used for other purposes, have been utilized to make the measurements. On a smaller scale, less expensive and less precise TIMS instrumentation has been used to pursue calcium metabolic studies (9).

Chemical separations and isotopic analysis capabilities have been established for a number of the elements of the periodic table. However, the inability to ionize many other important elements efficiently has posed a major problem and has provided opportunities for research in a critical area. Recent developments in resonance ionization, thermal formation of positive molecular ions and of negative ions on complex emitting surfaces, promise to expand the ability to form and measure ions at varying levels of sensitivity for nearly every element in the periodic table. These developments are expected to have a broad impact on the measurement of stable isotope ratios and concentrations of essential trace elements in humans, and to lead to studies of the bioavailability of these elements in metabolic processes.

In this review thermal ion formation processes will be summarized first and the general applicability of TIMS in inorganic elemental (mineral) analysis will be described. In contrast to this traditional (established) measurement technique, the new ionization technique, resonance ionization, will be introduced and the exciting prospects of this technique explicated. Furthermore, the mass spectrometry instrumentation and chemical separations used in stable isotope elemental analysis are summarized. The present and future roles of stable isotope measurements in nutrition will be assessed.

Thermal Ion Formation Processes

Thermal ionization from hot surfaces has been used extensively to produce positive ions for isotopic analysis. The efficiency of positive ion formation for selected alkali and other elements is controlled by the Saha-Langmuir expression:

$$n^{+}/n^{o} = A \exp^{[(W-IP)/kT]} \qquad (1)$$

where n^{o} = neutral atoms, n^{+} = positive ions, A = g_{+}/g_{o} = statistical weight ratio of ionic to atomic species leaving the surface, IP = ionization potential of the element being ionized, W = work function of the surface, k = Boltzmann constant and T = temperature in Kelvin. For alkali elements, the ionization efficiency can approach 100%, although

n^+/n^o ratios orders of magnitude lower are typical for most elements. For example, RIMS has been used recently to demonstrate that, at 1230K, an elemental iron deposit on a rhenium substrate produces an atomic vapor phase whose Fe^+/Fe^o ratio is $< 10^{-12}$ (11). In general, higher ionization efficiencies are predicted for substrates with higher work functions (W), elements with lower ionization potentials (IP) or for higher operating temperatures (T). In fact, for most real samples, Equation 1 is inadequate to predict the ionization efficiency. Samples may consist of multilayer mixtures of condensed phase compounds that vaporize to form molecules in addition to atoms (12-13). The work function of the surface can be altered by changing the orientation of the crystal (14), and by the formation of oxides (15). Volatile compounds containing elements with low-to-moderate ionization potentials have been dealt with by using a multiple filament thermal ion source (MFTIS). With this source, the volatile compound or element is vaporized from one surface (or two) at a lower temperature and ionized at a second, higher temperature surface (16). Using the conventional single or multiple filament ion source, techniques have been developed at NBS to measure isotope ratios for more than 25 elements at varying levels of sensitivity . About fifteen years ago, a method of enhancing the ionization efficiency was devised that employs the addition of silica gel and phosphoric acid to a rhenium substrate(17-18). This method has greatly enhanced the measurement sensitivity for numerous elements, and has revolutionized the ability to measure low level concentrations of lead and lead isotope ratios in natural matrices (19-20). The mechanism for this enhancement is not well understood; however, it is thought to be due either to an effective lowering of the analyte ionization potential, perhaps through an associative ionization mechanism, or to an increased ionization probability forced by diffusion through a heated matrix. More recently, the silica gel-phosphoric acid approach has been used for the formation of the arsenic monosulfide molecular ion, which has formed the basis for precise IDMS and isotope ratio measurements of sulfur (21). The elements determined by positive ion TIMS are summarized in Figure 1A (22).

Thermal ionization can also be used to form negative ions. Some elements with high ionization potentials that form positive ions inefficiently, such as the halides, can be analyzed using the negative ion formed in otherwise conventional thermal sources. Negative ion formation can be rationalized using a revised version of Equation 1, in which the proportion of negative ions is increased by lowering the work function of the emitting surface or by utilizing the high electron affinity of the element :

$$n^{-}/n^{o} = A \exp^{[(W-EA)/kT]} \quad (2)$$

where the parameters are analogous to Equation 1, except that the ionization potential of the element, IP, has been replaced by the elemental electron affinity, EA. Chlorine and bromine negative atomic ions have been used to determine the atomic weights of these elements and to measure their concentration by IDMS (23-25). The dramatically lowered surface work functions and enhanced negative ion emissions obtained with complex emitting surfaces (26-27) have led to the further development of these surfaces as efficient ionization sources. Lanthanum hexaboride and tantalum carbide have been used to coat metal surfaces to enhance negative ion formation, and have been key ingredients in the recently demonstrated sensitive measurements of iodine (28-29). Considerable progress in negative ion mass spectrometry has been realized recently by Heumann et al (30). A summary of the elements determined by negative ion formation and measurement is presented in Figure 1B. The method may be useful for the analysis of a number of transition metals and nonmetals, although development will be required to utilize the potential. A comprehensive evaluation of the applications of negative thermal ionization is beyond the scope of this paper. However, an extensive review of experimental methods and techniques of negative ion production with surface ionization has been published recently (31).

Techniques have been developed to overplate samples, via electrodeposition, on otherwise conventional substrates, such as rhenium(32). Overplating with rhenium and other transition and noble metals appears to provide three advantages: the ionization probability of the analyte is increased by forcing a diffusion of the atoms through a hot metal layer; the sample depletion with time is also lessened by the same diffusion mechanism; and

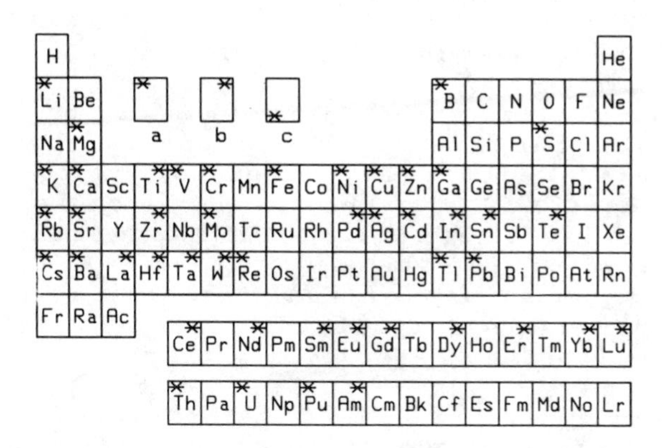

Figure 1A. Periodic table depicting utility of positive ion TIMS (mononuclidic elements have not been considered). a, Elements for which isotope dilution technology has been developed at NBS; b, additional elements for which isotope ratio or isotope dilution measurements have been reported (22); and c, elements for which positive ion TIMS is potentially applicable.

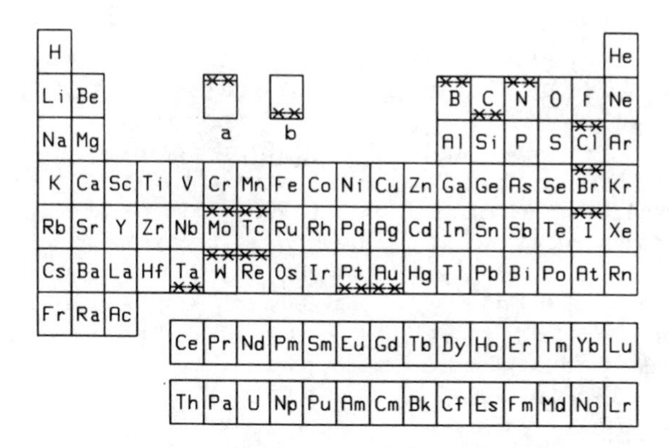

Figure 1B. Periodic table depicting utility of negative ion TIMS. a, Elements for which isotope ratio determinations have been reported; and b, negative thermal ions detected (30).

the need for chemical separation is diminished for very small samples.

The isotope ratios produced by thermal ionization are usually not representative of the actual isotope ratio in the sample being analyzed. Mass discrimination effects that occur in the thermal vaporization and ionization processes can induce isotope ratio measurement errors of up to several percent. Research into the origin of these effects during the last several years has successfully explained many of these 'anomalies'(12,13,33). Molecular vaporization of condensed phase compounds and elements can be explained through vaporization models developed by Kanno (12) and by Moore, Heald, and Filliben (13). In Figure 2, a typical calculation is illustrated for the vaporization of potassium compounds, but it is generically applicable to other elements and compounds as well. The magnitude of the mass discrimination effect for this system can be related directly to the chemical form in which the potassium vaporizes from the surface. Calculations of the equilibrium vapor composition for the potassium chloride system, at temperature and pressure conditions that prevail in the thermal ion source have been completed using techniques based on minimization of free energy (34). These calculations are summarized in Figure 3 for the potassium chloride system, for which requisite reliable thermodynamic data exist. Many avenues of indirect experimental isotopic analysis evidence exist to support these calculations (13), and the theoretical implications are considered to be generically applicable to many experimentally tractable inorganic compounds.

Solutions exist to deal with many of the mass discrimination effects of TIMS. Except for those elements that participate heavily in natural physico-chemical reactions, such as C, H, O and S, and those isotopes that contain stable isotope components, such as ^{87}Sr and ^{143}Nd, that are derived from the decay of long-lived radiogenic parents, most elements in nature are isotopically invariant. Thus, for many elements that have three or more isotopes, corrections can be made for the mass discrimination effects induced in the vaporization process by using internal normalization techniques (35). Techniques that utilize a double isotope spike provide a more elegant approach to correct mass discrimination effects for elements that have four or more isotopes (36-37).

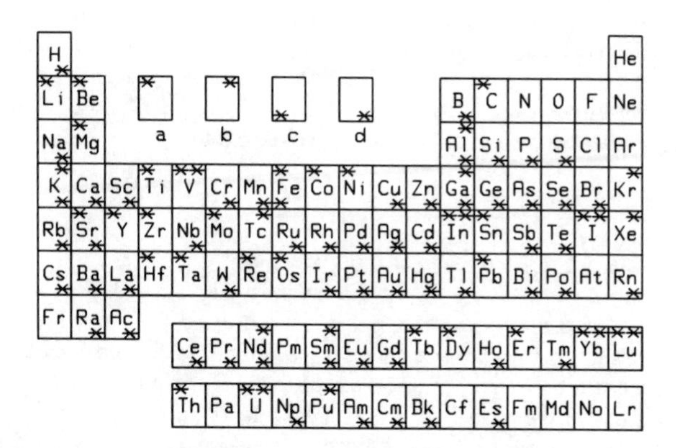

Figure 1C. Periodic table depicting utility of RIMS. a, Elements for which resonance ionization feasibility has been demonstrated at NBS using thermal atomization; b, elements for which ionization feasibility has been demonstrated in other laboratories using resonance ionization mass spectrometry (38-41, 44-52); c, elements for which isotope dilution RIMS have been achieved at NBS; and d, potentially applicable for resonance ionization via two and three photon processes (Schemes 1, 2, and 5 of Figure 4), using the RIMS system in Figure 6.

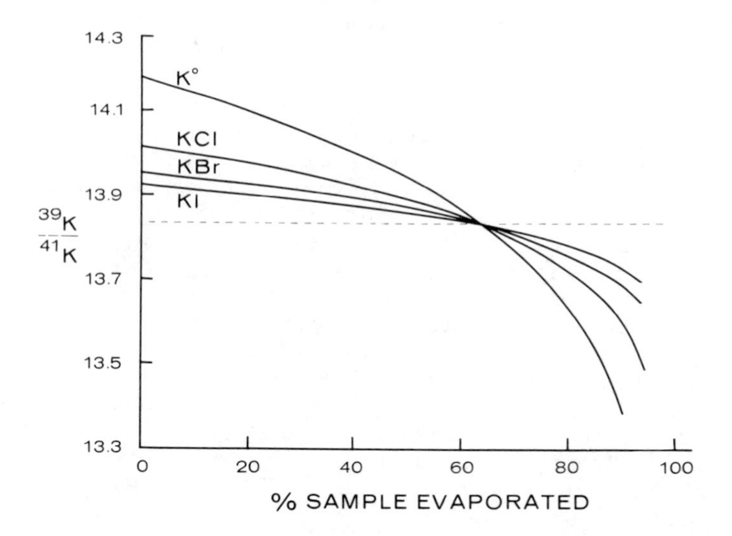

Figure 2. Isotopic fractionation curves computed from Kanno (12) for the vaporization of atomic potassium and potassium compounds. The absolute $^{39}K/^{41}K$ ratio (13.8566 $\pm$ 0.0063) is denoted by the dotted line.

Since these mass discrimination effects are largely a function of the vaporization process, the same discrimination and correction techniques would be applicable to the resonance ionization processes that utilize thermal vapor sources.

Resonance Ionization Processes

As the preceding discussion describes, thermal ionization from hot surfaces has limited scope. A new and comprehensive method of ionization is evolving that appears capable of ionizing nearly every element in the periodic table (11,38-42). Coupled with thermal vaporization processes and mass spectrometry, resonance ionization has already been used in our laboratory to demonstrate ionization feasibility for more than one-fourth of the elements in the periodic table (43). The current status of feasibility demonstration for atomic resonance ionization mass spectrometry in this and other laboratories is summarized in Figure 1C (39-41,44-52). This summary includes experiments performed with thermal and ion sputter-initiated techniques.

Resonance ionization requires the use of a laser to excite an electron to a specific atomic energy level, via one or more photons, and to eject the excited electron into the ionization continuum with an additional photon. Several schemes proposed for the process are summarized in Figure 4 (38). In our laboratory we have used a modified version of scheme 1, a two photon process in which two frequency doubled photons of equal wavelength are used. The first energy level is selected to be more than halfway to the ionization potential of the element, so that a second photon of the same wavelength completes the ionization. This simple process should be applicable to the ionization of more than fifty elements. Recent experimental work in our laboratory has indicated that schemes 1,2 and 5, using fundamental and frequency-doubled photons from a single laser, offer the simplest and most analytically comprehensive combination of the schemes. Special purpose, high selectivity applications may require more sophisticated optical approaches.

Two basic and unique advantages are implicit in the resonance ionization process: ionization efficiency and wavelength selectivity. Absorption cross-sections for excitation to the first energy

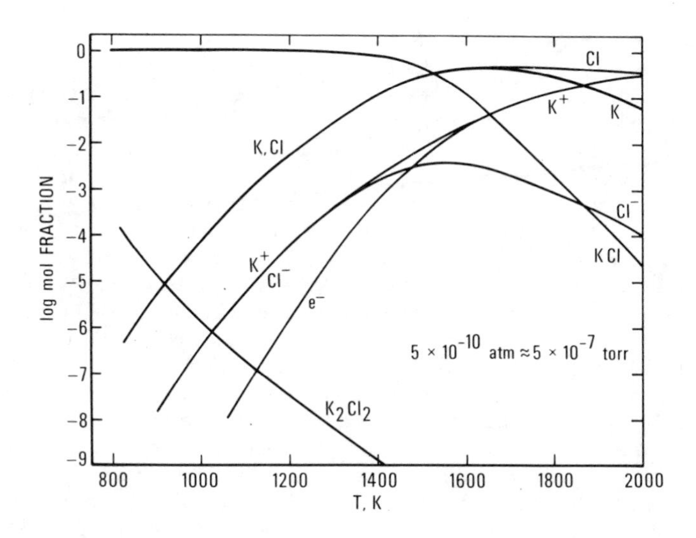

Figure 3. Vapor composition over condensed phase KCl as a function of the temperature range normally encountered in TIMS. These data were computed by Heald (34) using minimization of free energy techniques.

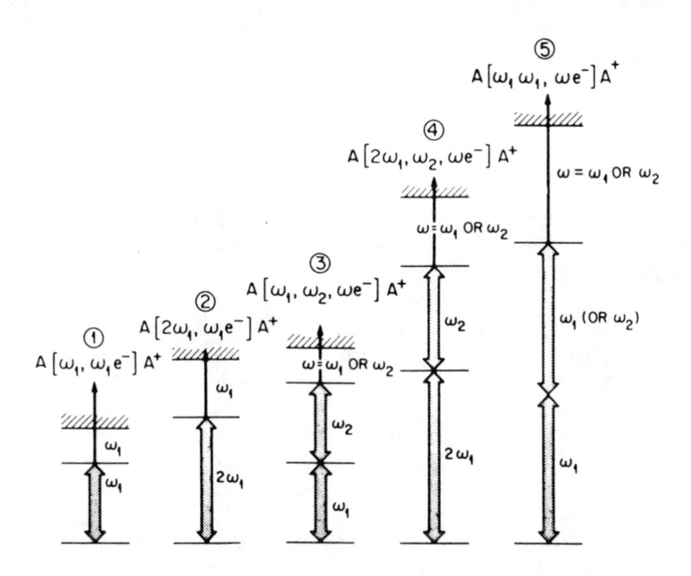

Figure 4. Classification of resonance ionization schemes for atomic ion formation (38). ω_1 and ω_2 denote photons of frequency 1 and 2, respectively; $2\omega_1$ denotes a photon that has been generated by frequency-doubling.

level are typically 10^{-12} cm^{-2}, and for the second step, 10^{-17} cm^{-2}. Photon fluxes available from commercial Nd:YAG pumped tunable dye laser systems are capable of saturating these cross-sections, such that Nxsigma > 1, where N= number of photons, and sigma= photon absorption cross section, which implies unit ionization efficiency; that is, every atom in the proper electronic state that passes through the laser beam will be ionized. Since a specific laser wavelength must be accessed to ionize an element, resonance ionization provides wavelength selectivity. Depending on the elemental atomic energy level distribution, it is possible to ionize elements selectively in the presence of other elements and molecules. Using narrower laser bandwidths, the additional level of isotopic ionization selectivity may be achievable for selected elements(53).

Thermal vapor sources used in TIMS can be directly applicable to the atomization required for resonance ionization. In general, direct atomization is possible from metallic deposits of elements on appropriate substrates. These deposits are achieved either through reduction in a hydrogen atmosphere to the metal or by in-situ reduction in a graphite slurry (11,42-43). These methods have been used to produce spectra for the four elements in Figure 5. Since more than eighty percent of the periodic table has a metallic character, this approach is broadly applicable. Moreover, since about thirty elements can be accessed in a given tunable dye frequency range, a potential multi-element analysis capability exists for mixed transition metals and nonmetals, including those elements important for nutritional bioavailability studies (54).

Although numerous laser systems have been applied to RIMS, none has proven to be more versatile than the Nd:YAG-pumped tunable dye system. A schematic of the RIMS system used in our laboratory is illustrated in Figure 6. The duty cycle of the laser ionization system is low compared to the continuous TIMS ion source; i.e., the combined thermal diffusion and laser repetition rate of 10 Hz (approximately 10 ns/ pulse) limits the ion production to an ON/OFF ratio of about 10^{-5}. For most trace element applications with nanogram and microgram size samples the duty cycle is not a measurement limitation. In fact, the use of a time-gated detection system can be a great advantage in improving the signal to noise ratio by factors that

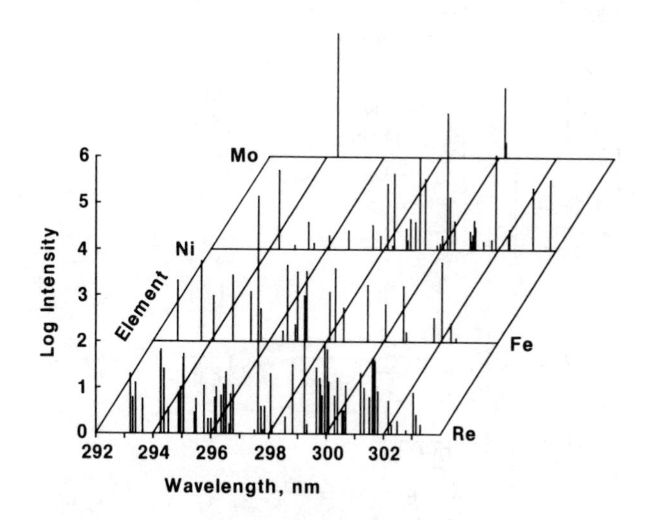

Figure 5. Resonance ionization spectra of the elements Re, Fe, Ni, and Mo. For each element the wavelength of the frequency-doubled dye output was scanned while focussing the mass spectrometer on the most abundant atomic ion of the element. Each wavelength dependent peak represents ion formation from a specific ground state or metastable atomic energy level (11).

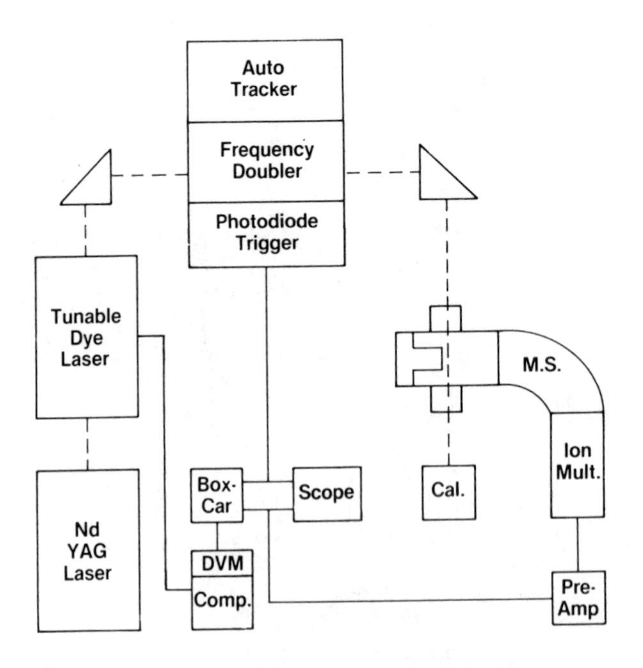

Figure 6. Block diagram of the experimental RIMS system, which consists of three basic components: a laser system capable of producing tunable ultraviolet radiation, a magnetic sector mass spectrometer with a suitably modified thermal atomization source, and a detection and measurement circuit capable of quantifying the pulsed ion currents produced in the experiment.

are commensurate with the reciprocal of the duty cycle: 10^5. Thus, continuously produced hydrocarbon backgrounds from a hot surface can be discriminated against and reduced to negligible levels. However, to approach the ultimate in absolute sensitivities a pulsed atom source has been developed to produce full-width, half maximum (FWHM) atom pulses < 1 ms, which potentially improves the overall duty cycle to > 10^{-3} (55). Improvements in this pulsed atom source are expected to reduce the FWHM substantially.

Ion Manipulation and Measurement

The most accurate and precise thermal ionization isotope ratio measurements have always been achieved with magnetic sector mass spectrometers, and this type of instrumentation has been used for isotope enrichment studies in nutrition with thermal ionization (8,10,56) or electron impact ionization (57). Less expensive alternates to this approach are quadrupole and time-of-flight mass spectrometers. Quadrupole mass spectrometers have been used extensively for gas chromatography-mass spectrometry (GC-MS), often via metal-chelate compounds, for isotope ratio measurements of nutritionally important elements (58-59). There has been a renewal of interest in time-of-flight mass spectrometers due largely to a resurgence in the use of pulsed ion sources, such as ^{252}Cf and laser desorption(60-61).

Although it may not be necessary to achieve precisions and accuracies at a level of 0.01-0.1% to measure stable isotope enrichments in nutritional and bioavailability studies for some applications, TIMS will probably be the method of choice for at least immediate and medium term applications, because of the breadth of applicability, ion transmission efficiency of the ion optics, and the attendant precisions and accuracies that are unavailable from other instruments. Precision measurement capabilities, for example, would be appropriate for applications to extended kinetic studies, for which high precisions must be available to resolve increasingly small isotope enrichments as a function of time. Isotope ratio precisions better than 10 ppm are available from modern instrumentation and inaccuracies less than 0.02% can be achieved.

Less expensive and less precise instrumentation can be used for meaningful isotope ratio measurements within biological sampling uncertainties. However, long-term measurement experience in related geological areas suggests a philosophy in which the 'best' measurements are always made. The levels of precision required to achieve existing nutrition and bioavailability objectives preclude significant breakthroughs possible with the resolution of increasingly smaller variations in isotope enrichments. Phenomena that are obscured by sampling variations may limit the scientific advances.

Chemical Separations

Chemical separations for isotope measurements at NBS have evolved along two basic thrusts: highly precise and accurate chemical stoichiometry oriented toward absolute isotope abundance measurements, and chemical purification techniques for the isotope dilution analysis of trace elements in a variety of matrices. The technology developed for each of these thrusts has reinforced the other. Accuracies of 0.02% in chemical assays with separated isotopes have produced a knowledge base for high accuracy handling of chemicals, solutions and ion exchange separations. Much of this knowledge has been directly transferable to high accuracy isotope dilution measurements with TIMS (3). Conversely, the need to develop separation techniques for diverse matrices, ranging from soils to bovine liver, has produced a number of innovative approaches to chemical separations which have in turn provided analytical techniques that have been employed in chemical stoichiometry for absolute abundance measurements. Chemical separation of the nutrition analyte from the matrix is an essential part of isotope dilution methodology (each separation is usually very matrix-specific), since a highly purified analyte is required to produce thermal ions efficiently and free of interferences. Cation and anion exchange separations have proved sufficiently versatile for most elements studied to date, and solvent extraction has been used for special cases such as chromium in biological materials(62). Electrodeposition of certain elements such as lead has been very effective as a final purification step prior to analysis with silica gel-phosphoric acid enhanced ionization procedures.

Element- and group-specific separations have been achieved for multi-element analyses with chelating materials (63). Using this approach, chemical separations can be tailored for certain groups of elements that are specifically oriented toward a given matrix or analysis technique. Thus, transition element groups from ocean water samples can be selectively freed of alkali and alkaline earth element contamination prior to analysis with neutron activation or x-ray fluorescence (64-65). Analogously, similar separation schemes should be directly applicable to multi-element RIMS analyses, for which transition and other metals can be aggregated by groups that are peculiar to RIMS measurement capabilities (43). Such a capability would open up the possibility for multielement stable isotope tracer studies with the attendant cost-effective ability to measure inter-element bioavailability distributions and interactions among exchangeable body pools. Automation of chemical separations is currently underway, which would add substantially to the economy of measurement.

Sample Contamination

Trace element concentration determinations at the part per million and lower levels with IDMS require rigorous attention to contamination of the sample. Sources of contamination that comprise the analytical blank have been documented by Murphy (66). These are principally derived from chemical reagents, separations media, particulate fallout and instrumental sources, such as impurities in the filament material used to form thermal ions. Lead concentrations in ocean waters and natural materials have been recorded with isotope dilution by paying strict attention to such sources of contamination (67).

Present and Potential Applications to Nutrition and Bioavailability Studies

Isotope dilution mass spectrometry can be used to determine element concentrations at trace and macro levels with high sensitivity and high accuracy. Several examples of sensitive and accurate measurements with IDMS are illustrated in Table I for foodstuffs and biological matrices (2,4,35,62,68). The accuracy of IDMS has been used to certify elemental concentrations in a wide variety of matrices, with concentrations ranging over eight orders of magnitude.

Table I. Trace and Macro Elemental Concentrations In Selected Matrices

Element (Nuclide)	Matrix	No. Samp.	Conc. / Error	Units	Ref.
K	Freeze-Dried Serum	7	44.092 +-0.050[a]	umol/g	4
			44.102 +-0.027[a]	umol/g	
K	Reconstituted Serum	7	3.5189 +-0.0140[a]	mmol/g	4
(^{238}U)	Bovine Liver (SRM 1577a)	6	704+-11.9[b]	pg/g	62
U,total			709	pg/g	
Cr	Brewer's Yeast (SRM 1569)	6	2.12 +-0.13[c]	ug/g	54
Ca	Bovine Serum	12	3.570 +-0.008[c]	meq/L	2
			4.294 +-0.009[c]	meq/L	
			5.032 +-0.011[c]	meq/L	
			5.733 +-0.012[c]	meq/L	
Mo	Ore Concentrates (SRM 333)	12	55.312 +-0.044[c]	wt%	35

[a] standard deviation
[b] 2 x standard deviation
[c] t x standard deviation

Collaborative efforts with experts in nutrition and bioavailability have been carried out at NBS and have resulted in the use of TIMS for the measurement of stable isotope tracers in adults and newborns(8,10). Typical data for these efforts are summarized in Tables II and III for Ca and Zn isotopes. For each of these elements, optimum isotope ratio precisions of about 0.1% (R.S.D.)were achieved, which permitted the resolution of extended term kinetic effects that would be impossible to achieve with less accurate and precise mass spectrometry.

The calcium isotope ratio data of Table IIA were obtained using a triple filament thermal ionization procedure (2) that was modified to permit operation of the ionizing filament at a higher temperature

(1600°C vs. 1400°C). This is believed to preclude the deposition and re-emission of atomic calcium from the ionizing filament, and yields a lower observed $^{40}Ca/^{44}Ca$ ratio of 46.3469 compared to the earlier value of 46.4800 (2). The observed isotope ratios of Table IIA were normalized to $^{40}Ca/^{44}Ca = 46.4800$ to retain consistency with earlier measurements, and were also normalized to the average $^{42}Ca/^{44}Ca$ ratio to demonstrate the effectiveness of the proportional-to-mass difference fractionation correction. The value of internal normalization (35) can be observed by intercomparing Tables IIA and IIB. The data of Table IIB were acquired by measuring the calcium isotope ratios in serum samples taken at successive intervals following injection of a newborn human with ^{46}Ca and during continuous oral administration of ^{48}Ca. The ^{42}Ca, ^{43}Ca and ^{44}Ca contained in the ^{46}Ca and ^{48}Ca separated isotopes make negligible (<< 0.1%) contributions to the corresponding mass positions of the natural calcium in the serum sample (10). This permits normalization of the observed ratios to the average observed $^{42}Ca/^{44}Ca$ ratio and provides a correction for inter-analysis variations that might be as much as one percent. The $^{43}Ca/^{44}Ca$ serves as an additional cross-check, and when corrected to the same $^{42}Ca/^{44}Ca$ ratio for SRM 915 (Table IIA), the average $^{43}Ca/^{44}Ca$ ratio for the standard and samples agree within about 0.1%.

Table IIA. Isotopic Analyses of SRM 915, $CaCO_3$[a] (Relative to ^{44}Ca)

	^{40}Ca	^{42}Ca	^{43}Ca	^{46}Ca	^{48}Ca
Ave	46.3469	0.30941	0.064570	0.001506	0.090088
S.D.	0.0341	0.00022	0.000094	0.000029[b]	0.000163
Normalized to $^{40}Ca/^{44}Ca$= 46.4800 :					
	46.4800	0.30987	0.064616	0.001504	0.089831
S.D.	----	0.00011	0.000070	0.000029[b]	0.000082
Normalized to $^{42}Ca/^{44}Ca$ = 0.309411 :					
	46.3467	0.30941	0.064570	0.001506	0.090088
S.D.	0.0341	----	0.000076	0.000029[b]	0.000080

[a] (seven analyses)
[b] estimated uncertainty

Table IIB. Calcium Isotope Ratios in Serum[c,d] (Relative to ^{44}Ca)

Samp.	Post-Injec. t,min	^{48}Ca	^{46}Ca	^{43}Ca	^{42}Ca
1	15	0.092214	0.02850	0.06448	0.30891
2	30	0.093538	0.02558	0.06439	----
3	60	0.097412	0.01970	0.06442	----
4	720	0.13321	0.00482	0.06445	----
5	1440	0.13982	0.00299	0.06451	----
			Ave.	0.06445	
			S.D.	0.00005	

[c] Normalized to $^{42}Ca/^{44}Ca = 0.30891$
[d] Data used to compute a portion of the dilution curves of Ref. 10

Table IIIA. Data Comparing Precision of IRMS[a] and RNAA[b] for $^{70}Zn/^{68}Zn$ Ratios in Unenriched Samples[c]

Method	No. Determs.	Ave.	S.D.
IRMS	5	0.03247	0.00004
RNAA	6	0.00824	0.00068

Table IIIB. Comparative Isotope Ratio Data for RNAA and IRMS [c]

Time Post-admin.,min.	$(PA)_{386}/(PA)_{439}$	$(^{70}Zn/^{68}Zn)_{IRMS}$	k^d
Unenriched	0.00824+-0.00068	0.03247+-0.00004 (0.13%)	3.94
15	0.00956	0.03817	3.99
30	0.01001	0.04934	4.93
45	0.0124	0.05659	4.56
60	0.0145	0.06193	4.27
75	0.0162	0.06501	4.01
90	0.0163	0.07067	4.33
120	0.0164	0.07348	4.48
150	0.0162	0.07339	4.53
180	0.0144	0.06697	4.65
210	0.0144	0.07120	4.94
240	0.0172	0.07378	4.29
270	0.0161	0.07085	4.40
330	--	--	--
390	0.0128	0.05916	4.62
450	0.0131	0.06007	4.59
		Ave.	4.45+-0.30 (6.7%)

[a] IRMS = Isotope Ratio Mass Spectrometry

[b] RNAA = Radiochemical Neutron Activation Analysis

[c] Reference 8

[d]
$$k = \frac{(^{70}Zn/^{68}Zn)_{IRMS}}{(PA_{386}/PA_{439})_{RNAA}}$$

The ionization efficiency and selectivity of RIMS can be utilized for the analysis of ultra-small samples at atom-counting sensitivity levels. Simplified chemical separations may also result from the ability to analyze atoms selectively in the presence of other matrix components, within limitations that will be determined by ongoing research. Combined with the possibility of multielement analysis, RIMS may become a cost-effective way of providing multi-element isotope tracer information in nutrition studies. Most recently, RIMS has been used to measure iron concentrations in trace element in water and human serum SRMs, and a calibration of the RIMS system with iron isotopes has been achieved at precisions and accuracies of 1-2% over an isotope ratio range of nearly 1000 (69). The potential sensitivity of RIMS is illustrated by the observation of less than 10^{10} atoms of lead, and preliminary indications suggest that measurements of substantially less than 10^5 atoms will be achievable (70). The availability of these sensitivities and accuracies makes it conceivable that stable isotope tracer metal concentration studies can be extended to the exploration of the most minute partitioning levels in human and other biological samples. The emergence of biotechnology as a rapidly evolving trend in science may also be abetted by the application of RIMS. For example, the ability to examine the role of ultra-trace metal transport among cellular organisms and within biological systems would permit a better understanding of the function of essential nutrients and of optimal metallo-organic compound selection to maximize their bioavailability. Similarly, selective multi-photon resonance ionization of organic and metallo-organic compounds may provide a sensitive and broadly applicable method for the characterization of species that are important in biotechnology development. The ability to examine iron isotope ratios, one red blood cell at a time, may be just around the analytical corner.

Acknowledgments

The author appreciates critical review and suggestions by his colleagues, I.L. Barnes, J.D. Fassett and J.C. Travis. In order to describe adequately experimental procedures, it is necessary to identify commercial products by manufacturer's name or label. In no instance does such identification imply endorsement by the National Bureau of Standards nor does it imply that the particular products or equipment are necessarily the best available for that purpose.

Literature Cited

1. Thomson, J.J. "Rays of Positive Electricity"; Longmans, Green and Co., London, 1913; 2.

2. Moore, L.J.; Machlan, L.A. Anal. Chem. 1972, 44, 2291-96
3. Cali, J.P.; Mandel,J.; Moore,L.J.; Young, D.S."A Reference Method for the Determination of Calcium in Serum,"National Bureau of Standards Special Publication 260-36, 1972.
4. Gramlich,J.W.; Machlan,L.A.; Brletic,K.A.; Kelly, W.R. Clinical Chemistry 1982, 28, 1309-13
5. NBS Certificate of Analysis, Human Serum SRM 909
6. Janghorbani,M; Young,V.R. in "Advances in Nutritional Research,"H.H. Draper, ed., Vol. 3 Plenum Press: New York, 1980, Chapter 5
7. Janghorbani,M.; Ting,B.T.G.; Young,V.R. Am. J. Clin. Nutr. 1981, 34, 2816-30
8. Janghorbani,M.; Young,V.R.; Gramlich,J.W.; Machlan,L.A. Clinica Chim. Acta 1981, 114, 163-71
9. Yergey,A.L.; Vieira,N.E.; Hansen,J.W. Anal. Chem. 1980, 53, 1181-14
10. Moore, L.J.; Machlan, L.A.; Lim, M.O.; Yergey, A.L.; Hansen, J.W. "Dynamics of Calcium Metabolism in Infancy and Childhood I. Methodology and Quantification in the Newborn," submitted for publication
11. Fassett, J.D.; Moore, L.J.; Travis, J.C.; Lytle, F.E. Int. J. Mass Spectrom. and Ion Processes, 1983, 201-216
12. Kanno,H. Bull. Chem. Soc. Jpn. 1971, 44, 1808
13. Moore, L.J.; Heald, E.F.; Filliben, J.J. Adv. Mass Spectrom. 1978, 7, 448-74
14. McHugh, J.A. Int. J. Mass Spectrom. Ion Phys., 1969, 3, 267
15. Valyi, L. "Atom and Ion Sources"; John Wiley & Sons and Akademiai Kiado, Budapest, Hungary, 1977.
16. Inghram, M.G.; Chupka, P. Rev. Sci. Instrum., 1953, 24, 518
17. Akishin,P.A.; Nikitin, O.T.; Panchenkov Geochemistry(USSR), 1957, 500.
18. Cameron, A.E.; Smith, D.H.; Walker, R.L. Anal. Chem. 1969, 41, 525
19. Barnes, I.L.; Murphy, T.J.; Gramlich, J.W.; Shields, W.R. Anal. Chem. 1973, 45, 1881-84
20. Tera, F.; Wasserburg, G.J. Anal. Chem. 1975, 47, 2214-20
21. Paulsen, P.J. and Kelly, W.R. "Determination of Sulfur as Arsenic Monosulfide Ion by Isotope

Dilution Thermal Ionization Mass Spectrometry," Anal. Chem., in press

22. DeBievre, P.; Gallet, M.; Holden, N.E.; Barnes, I.L. "Isotopic Abundances and Atomic Weights of the Elements: An Evaluated Compilation," J. Phys. and Chem. Ref. Data, in press
23. Shields, W.R.; Murphy, T.J.; Garner, E.L.; Dibeler, V.H. J. Am. Chem. Soc. 1962, 84,1519
24. Catanzaro, E.J.; Murphy, T.J.; Garner, E.L.; Shields, W.R. J. Res. Nat. Bur. Stand.(U.S.) (Phys. and Chem.), 1964, 68A, 593
25. Garner, E.L.; Machlan, L.A.; Gramlich, J.W.; Moore, L.J.; Murphy, T.J.; Barnes, I.L., Proc. 7th IMR Symposium, Accuracy in Trace Analysis: Sampling, Sample Handling and Analysis; National Bureau of Standards Special Publication 422, 1976, p. 951.
26. Blais, J.C.; Bolbach, G. Int. J. Mass Spectrom. Ion Phys., 1977, 24, 413-27
27. Delmore, J.E.; Int. J. Mass Spectrom. Ion Phys., 1982, 43 , 273-281
28. Rankin, R.A.; Nielsen, R.A.; Hohorst, F.A.; Filby, E.E.; Emel, W.A. Proc. 31st Ann. Conf. on Mass Spectrom. and Allied Topics 1983, p. 845
29. Stoffels, J.J. Proc. 30th Ann. Conf. on Mass Spectrom. and Allied Topics 1982, p. 329.
30. Heumann, K.G.; Kastenmayer, P.; Schindlmeier, W.; Unger, M.; Zeininger, H.Proc. 31st Ann. Conf. on Mass Spectrom. and Allied Topics, Boston,MA, pp. 581-2
31. Kawano,H.; Page,F.M. Int. J. Mass Spectrom. and Ion Phys., 1983, 50,1-129
32. Perrin, R.E.; Rokop, D.J.; Cappis, J.H.; Shields, W.R. Proc. 29th Ann. Conf. on Mass Spectrom. and Allied Topics,1981, Minneapolis, MN , pp. 422-23
33. Habfast, K. Int. J. Mass Spectrom. Ion. Phys., 1983, 51, 165-189
34. Heald, E.F. Proc. 25th Ann. Conf. on Mass Spectrom. and Allied Topics, Washington, D.C., 1977, p. 509.
35. Moore, L.J.; Machlan, L.A.; Shields, W.R.; Garner, E.L. Anal.Chem. 1974, 46, 1082
36. Dietz,L.A., Pachucki, C.F., and Land, G.A., Anal. Chem., 1962, 34, 709-10
37. Dodson, M.H.; J. Sci. Instrum., 1963, 40, 289
38. Hurst, G.S.; Payne, M.G.; Kramer, S.D.; Young, J.P. Rev. Mod. Phys., 1979, 51, 767

39. Beekman, D.W.; Callcott, T.A.; Kramer, S.D.; Arakawa, E.T.; Hurst, G.S. Int. J. Mass Spectrom. Ion Phys., 1980, 34, 89-97
40. Donohue, D.L.; Young, J.P.; Smith, D.H. Int. J. Mass Spectrom. Ion Phys., 1982, 43, 293-307
41. Miller, C.M.; Nogar, N.S.; Gancarz, A.J.; Shields, W.R. Anal. Chem., 1982, 54, 2377-78
42. Fassett, J.D.; Travis, J.C.; Moore, L.J.; Lytle, F.E. Anal. Chem., 1983, 55, 765-70
43. Moore, L.J.; Fassett, J.D.; Travis, J.C. "Systematics of Multielement Analysis Using Resonance Ionization Mass Spectrometry and Thermal Atomization," in preparation
44. Winograd, N.; Baxter, J.P.; Kimock, F.M. Chem. Phys. Lett., 1983, 88, 581-84
45. Donohue, D.L.; Young, J.P. Anal. Chem., 1983, 55, 378
46. Balooch, M.; Olander, D.R. Int. J. Mass Spectrom. Ion Phys., 1983, 51, 155-64
47. Young, J.P.; Donohue, D.L. Anal. Chem., 1983, 55, 88
48. Donohue, D.L.; Smith, D.H.; Young, J.P.; McKown, H.S.; Pritchard, C.A. Anal. Chem., 1984, 56, 379-81
49. Parks, J.E.; Schmitt, H.W.; Hurst, G.S.; Fairbank, W.M., Jr. Proc. of SPIE-The Int. Soc. for Optical Engineering: Symposium on Laser-Based Ultrasensitive Spectroscopy and Detection V, 1983, 426, pp. 32-39.
50. Chen, C.H.; Hurst, G.S. Proc. of SPIE-The Int. Soc. for Optical Engineering: Symposium on Laser-Based Spectroscopy and Detection V, 1983, 426, pp. 2-7
51. Miller, C.M.; Nogar, N.S.; Downey, S.W. Proc. of SPIE-The Int. Soc. for Optical Engineering: Symposium on Laser-Based Ultrasensitive Spectroscopy and Detection V, 1983, 426, pp. 8-9
52. Kimock, F.M.; Baxter, J.P.; Winograd, N. Surf. Sci., 1983, 124, L41
53. Lucatorto, T.B.; Clark, C.W.; Moore, L.J. "Ultrasensitive Mass Spectrometry Based on Two-Photon, Sub-Doppler Resonance Ionization," Optics Communications, 1984, 48, 406-410.
54. Mertz, W. Science, 1981, 213, 1332-1338
55. Fassett, J.D.; Moore, L.J.; Shideler, R.W.; Travis, J.C. Anal. Chem., 1984, 56, 203-6
56. Turnlund, J.R.; Michel, M.C.; Keyes, W.R.; King, J.C.; Margen, S. Am. J. Clin. Nutr. 1982, 35, 1033

57. Johnson, P.E. J. Nutr. 1982, 112, 1414
58. Hachey, D.L.; Blais,J.C.; Klein, P.D. Anal. Chem. 1980, 52, 1131-35
59. Swanson, C.A.; Reamer, D.C.; Veillon, C.; Levander, O.A. J. Nutr. 1983, 113, 793
60. Macfarlane, R.D. Biomedical Mass Spectrom. 1981, 8,449
61. Cotter, R.J. Anal. Chem. 1981, 53, 719
62. Dunstan, L.P.; Garner, E.L. Proceedings of Trace Substances in Environmental Health-XI, 1977, pp. 334-37
63. Kingston, H.M.; Barnes, I.L.; Brady, T.J.; Rains, T.C.; Champ,M.A. Anal. Chem. 1978, 50, 2064-70
64. Kingston, H.M.; Pella, P.A. Anal. Chem. 1981, 53, 223-27
65. Greenberg, R.; Kingston, H.M. Anal. Chem. 1983, 55, 1160-65
66. Murphy, T.J. Proc. 7th IMR Symposium, Accuracy in Trace Analysis: Sampling, Sample Handling and Analysis; National Bureau of Standards Special Publication 422, 1976, pp. 509-39
67. Settle,D.M.; Patterson, C.C. Science, 1980, 207,1167-1176
68. Kelly, W.R.; Fassett, J.D. Anal. Chem., 1983, 55, 1040-44
69. Fassett, J.D.; Powell, L.J.; Moore, L.J., "The Determination of Iron in Serum and Water by Resonance Ionization Isotope Dilution Mass Spectrometry," submitted for publication.
70. Moore, L.J., unpublished data

RECEIVED April 23, 1984

2

Calcium Metabolism

Studied with Stable Isotopic Tracers

A. L. YERGEY and N. E. VIEIRA—Laboratory of Theoretical and Physical Biology, National Institute of Child Health and Human Development, Bethesda, MD 20205

D. G. COVELL—Laboratory of Theoretical Biology, National Cancer Institute, Bethesda, MD 20205

J. W. HANSEN[1]—Neonatal and Pediatric Medicine Branch, National Institute of Child Health and Human Development, Bethesda, MD 20205

Physiologically important calcium flow rates can be determined using calcium isotopic tracers without resorting to classical metabolic balance methodology. Calcium stable isotopic tracers are sufficiently benign to permit their use in studies of calcium metabolism in children. The measurement techniques and clinical protocol for our studies of skeletal development in children are described. Partial results of studies of normal newborns and children with abnormalities of calcium metabolism are given.

Over the past several years we have been using highly enriched calcium isotopes as tracers for the elucidation of calcium metabolism in children. Normal growth and development of the skeletal system results in the accumulation of about 1000g of elemental calcium over about 15 years. The average rate of this calcium accretion is about 180 mg/day, but it is apparent that the accretion rate cannot be uniform. The rate must be much higher than average during a number of periods of marked growth which would then be bracketed by plateaus of slower development. It is expected that, on the average, normal children must be in positive calcium balance during the entire growth period, but during periods of rapid skeletal growth, calcium balance must be strikingly positive. There are, however, almost no direct measurements available to show the manner in which calcium is distributed internally, i.e. calcium kinetic parameters. Such measurement of these should lead to a more definitive determination of bone accretion rate than is possible from classical balance studies. The most likely reason for this dearth of kinetic information is that, until recently, studies of calcium metabolism generally required experiments that exposed

[1]Current address: Meade Johnson Corp., Evansville, IN 47711

the subject to radioactive isotopic tracers. The use of radiolabled materials in normal children represents an unacceptable ethical situation. Our development of a rapid analytical methodology that permits the use stable calcium isotopes as tracers (1),permits studying calcium metabolism with virtually no risk to the subject.

In order to carry out such studies successfully, a number of complex tasks must each be executed at a high level of competnce. Not only are the costs of stable isotopes high for such studies, and the commitment of time by the group undertaking each study substantial, but any extended study involving humans, especially children, must be justified only with carefully designed and executed procedures. The procedures that we have developed and use for such studies fall into three major catagories, isotopic analysis, clinical protocol and data analysis, which we will discuss further.

Isotopic Analysis

Thermal ionization mass spectrometry is used for the analysis of calcium tracers in preference to other potential analytical methodologies. Thermal ionization is a direct ionization method that produces intense ions at the isotopic masses and virtually nowhere else in the mass spectrum. It is superior to potential GC/MS methods employing chelates of calcium since all of these methods involve correcting observed ion intensities for the presence of elements in the ligand, even though such methods give reasonable isotope ratios (2). Virtually any mass spectrometric analysis is chosen over neutron activation analysis (NAA) because of the measurement of tracer abundance is done by a ratio technique in mass spectrometry versus the absolute measurement required by NAA. As a consequence of the ratio measurements, the sample workup is simpler and the accuracy possible is greater for mass spectrometric measurements compared to NAA. While thermal ionization is the preferred ionization method at the present time, ionization techniques that may become important in the future include inductively coupled plasma (ICP), fast atom bombardment (FAB) and laser desorption. While all of these methods have the potential for greater sensitivity than thermal ionization, they have not been used for the high accuracy/precision analysis of biological samples.

We have developed a thermal ionization method for use in a standard quadrupole mass spectrometer (1). The method uses a modified solids inlet probe, Figure 1, in conjunction with replacable filament assemblies. The rhenium filaments are coated with calcium salts precipitated from biological materials in a basic ammonium oxalate solution. Calcium is precipitated directly from urine and serum first made basic with ammonium hydroxide; fecal samples and aliquots of diet are homogenized with a 9:1 water/nitric acid mixture, centrifuged, made basic and then precipitated.The precipitates are conditioned in a steam bath and centrifuged. After drying, precipitates are heated at 500 °C to convert the oxaltes to carbonate/oxide salts, then disolved in 3% ultrapure nitric acid and placed on a filament. Filaments are degassed in a vacuum bell jar prior to being coated, and are dried under a heat lamp after coating. Dried filaments are placed on the tip of the probe, inserted into the vacuum system of the mass spectrometer, and finally positioned in the source region. The filaments are heated directly while on the probe, and ions are produced solely from this heating; no other ionization means are used. The response of the m/z 40 intensity from calcium as a function of time is shown in Figure 2. Ion intensities are measured when the ion signal has stabilized to the plateau region of the curve using a dedicated microprocessor controlled data acquisition device (3). Natural abundance isotope ratios obtained with the method are shown in Table I, and compared to the best measured values (4).

Table I. Calcium Isotope Ratios

	Observed	Best Measured
44/42	3.203 ± 0.006	3.203 ± 0.001
46/42	0.0050 ± 0.00003	0.00486 ± 0.00003
48/42	0.284 ± 0.001	0.284 ± 0.0008

While the agreement of our natural abundance measurements with the best measured values are quite good, the average relative standard deviation of measurements by our method is about 0.2% compared to about 0.1% for the best measured values. It is important to note, however, that our measurements are well within error limits that might reasonably be expected from measurements of a biological system. Our method, while a modification of that used to produce the best measured values (4), has introduced some simplifications to the sample workup that speed sample processing.

Thermal ionization has been used to determine isotopic abundance of virtually all the elements. We have recently extnded our own capability in this direction by adapting the silica gel/phosphoric acid filament coating technique (5) to our system. Five μl of a fine silica gel suspension is placed on a filament. Five μl of the analyte ion solution is coated, dried then coated with 2 μl of a 0.7N phosphoric acid solution and heated until dry again. The analysis is performed in a similar manner as before, except that the signal is more transient and somewhat less intense than the calcium analysis. With this approach, however, we have made natural abundance isotope ratio measurements on zinc, copper, and magnesium. Table II shows our measurements compared to the accepted values, shown in parenthesis, for these elements. The isotope used as reference

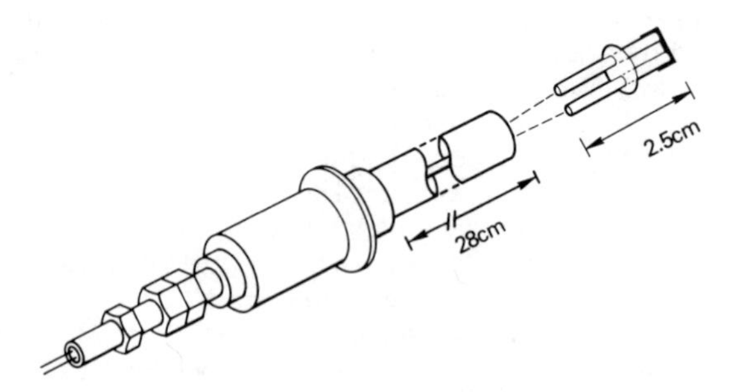

Figure 1. Thermal ionization solids inlet probe and disposable filament. Note two different scales. (Reproduced from Ref. 1. Copyright 1980, American Chemical Society.)

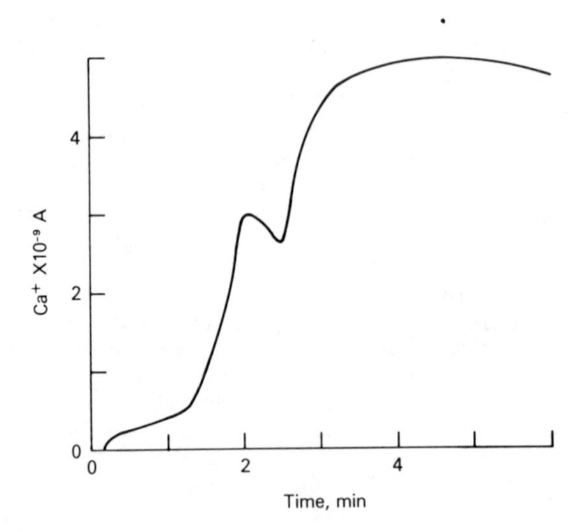

Figure 2. Evolution of Ca+ from thermal ionization filament. (Reproduced from Ref. 1. Copyright 1980, American Chemical Society.)

for the ratios of each element is shown by a dashed line in the Table.

Table II. Natural Abundance Ratios of Several Metals

	64	66	67	68
Zn	1.766	--	0.152	0.220
	(1.758)		(0.148)	(0.223)
	63	65		
Cu	2.077	--		
	(2.235)			
	24	25	26	
Mg	7.858	--	1.066	
	(7.769)		(1.103)	
(accepted value)				

Table II shows quite reasonable results for natural abundance isotope ratios, it must be noted that Cr, Fe and Mg are frequently observed as interferences. Heating a clean, but unloaded filament gives rise to Cr and Fe ions implying that these two elements are contaminants of the rhenium filament ribbon. Heating a filament loaded with silica gel and phosphoric acid gives rise to Mg ions as well as those from Cr and Fe while only Cr and Fe ions are seen when silica gel alone is on the filament. This implies a Mg contamination of the acid. Finally, inhibition of one metal's emmission by another has been seen previously (5). The previously reported inhibition of iron by zinc, and of cadmium by iron is seen in this work when the metals of Table II are contaminated with the appropriate ions.

Despite these difficulties, the use of the silica gel technique for the solid probe/quadrupole mass spectrometer system holds promise for the analysis of some nutritionally important metals. Both zinc and copper have been extracted successfully from serum, and zinc has also been extracted from urine and feces, by using an anion exchange purification. Biorad AGlX8 (100-200 mesh, chloride form) anion exchange resin has been used to separate copper and zinc from acidic solutions (6). We have adapted this method to the separation of these two metals from acidic solutions of AAS standards, urine, serum and deproteinated fecal homogenate by elution with sucessively dilute acid solutions. Recovery of an isotopic spike and subsequent mass spectral analysis has been demonstrated with a ^{70}Zn spike added to 1ml aliquots of a Fisher Certified AA Standard (zinc concentration = 1mg/ml).Results of this experiment are shown in Table III.

Table III. Zn Isotope Spike Recovery

Spike Added	Enrichment Observed	Recovery
25%	20.2%	81%
50	43.8	87
75	66.2	88
100	92.4	92

Clinical Protocol/ Data Acquisition

The clinical protocol, which includes the data acquisiton portion of a study, must be designed with some expectation of the results; these results are used to shape the protocol further. The hypothesis/model that was used as the analysis portion of this study was used to simulate the experimental measurements and to design the initial studies. Analysis of results was done by referring to this model. The results were used to modify the protocol for the second studies. The model is discussed below in the data analysis section.

Clinical protocols are designed to maximize the amount of data that can be obtained in a particular study while minimizing the risk to the subject. The situation is complicated since our major interests are the investigation of calcium kinetics in children. The protocol must be designed to be benign, yet give the required kinetic data. Because the subjects are minors, informed parental consent must be obtained in addition to the assent of the subject.

Two tracers are given simultaneously, one orally the other intravenously. The isotopes that are candidates for administration are shown in Table IV. The isotopes that are actually used are chosen for the optimum balance of cost and availability. The dose levels are most critical for the i.v. isotope since the mass of it that is injected must be great enough to provide a perturbation of the natural abundance level of that isotope for the entire time period of the study while not perturbing the plasma calcium homeostasis. The sensitivity of the measuring technique is critical to the isotope selection since the ability to detect small changes in natural abundance levels permits the use of smaller i.v. doses. For our measurement methodology, the dose levels shown in Table IV yield initial elevations in the natural abundance levels of the 46Ca and 48Ca of about 10 fold over natural levels and 5 fold elevations over natural levels for the 42Ca. The 44Ca cannot be used as an i.v. tracer because its natural level is too great to yield adequate elevations of the natural levels without perturbing plasma calcium levels.

Table IV. Calcium Isotope Doses

	42Ca	44Ca	46Ca	48Ca
Oral Dose (mg/kg)	0.64	2.5	--	0.22
i.v. Dose (mg/kg)	1.2	--	0.03	0.73
Cost ($/mg)	16.70	3.60	1492.00	167.00
Isotopic Purity	98.0	98.6	34.9	97.7

Participants in our recent studies were normal children between the ages of 6 and 15 who were within 20% of normal weight for their age and height. They were brought into the hospital with a stabilized daily calcium intake. After acclimatizing to the hospital routine, the isotope was injected in the morning of the first day, and the other isotope was given orally in their breakfast milk. Blood samples were drawn periodically during the first 12 hours, at .083,.167,.25,.5,1,2,4,8,12 hrs; urine samples were collected for the next 2 wks, as 8 hr. pooled samples for the first 6 days and as 12 hr. pooled samples thereafter. Two non absorbed fecal markers were used in this work. The first is a colored marker given daily with breakfast for three days beginning with the day of isotope administration. Polyethyleneglycol (PEG) is given with each meal for the same three days. Stool samples are collected for three days after the first colored marker is passed. PEG is used as a recovery standard, and is analyzed by a modification of the standard nephalometric method (7).

Results of isotope ratio measurements from an earlier study, that used a similar protocol to the one just described, are shown in Figure 3. These curves show plasma isotope levels for both the i.v.oral tracer, labelled as PTA and CTA respectively. It is generally accepted that isotopic enrichment, atom percent excess, of urine reflects that of plasma, and after the initial period of rapid mixing, urinary atom percent excess is used in lieu of plasma measurements. The curves drawn through the data are those generated using the proposed model for calcium kinetics.

Data Analysis/Modeling

The SAAM (Simulation, Analysis And Modeling) computer program developed by Berman and Weiss (8) was used to analyze the isotope dilution and balance data. The observed time dependent dilution of both calcium tracers in plasma, as reflected in urine at longer times, coupled with their cumulative appearance in urine and feces is used to calculate kinetic parameters of the model. The balance data are then used to calculate the steady state fluxes of the system. It is assumed that the tracers distribute themselves throughout the subject's bodies in physiological pools that are represented by the mathematical analogy of compartments in the model. It should be noted however, that there is no physiological identity ascribed to any of the mathematical compartments. The SAAM program permits direct input of observed data. A proposed model is also entered into the program, and a statistical fit between the observations and the hypothetical system is calculated. The quality of the fit is a basis for assessing the quality of the model. The minimum number of internal compartments that can be used to provide an acceptable fit is the basic premise for establishing the number of internal compartments. The data are fit to a model that is the equivalent of the classical sum of exponential terms, but uses the deconvolution methods developed by Berman and Weiss.

The basic model used in this work is shown in Figure 4. The three compartment serial model shown uses three internal compartments that each mix more slowly in progression. These three represent the overall rapidly exchangeable calcium pool with compartment 1 being assumed to include plasma and extracellular calcium. Bone deposition is assumed to occur from the third compartment of the model. This third compartment is presumed to be at least partially associated with the rapidly exchanging calcium deposits at the surface of bone. Bone resorption is assumed to place calcium back into the first compartment. This model is a simpler version of the four internal compartment system proposed by Neer et al. (9) and used by a number of others (10,11). The present work employs a simpler model than that used in the Neer, et al work because the time resolution of the early portion of our studies is such that only one compartement is required where the earlier workers used two. This is a consequence of slow sampling after the isotope injection. Further assumptions that are made in this modelling are that: 1) the subject is in a steady state with respect to calcium metabolism during the time period of the study; 2) the injected isotope mixes instantly after injection; 3) gastrointestinal absorption occurs continuously during the experiment; 4) tracer is not resorbed from bone during the experiment. These assumptions are simplifying assumptions that serve to make the model work, and are generally reasonable approximations.

Results from several studies are presented below in Tables VI and VII in terms of calcium flow rates. The symbols used to express the calcium flow rates are defined in Table V and shown schematically in Figure 4. Data from a newborn subject (12) in some earlier studies are shown in Table VI in comparison to data from an adult (13). All values are expressed per kg body weight per hour. The newborn subject shown in Table VI exhibits some interesting contrasts with the typical adult shown there; these contrasts are consistent with other subjects of this study (12).

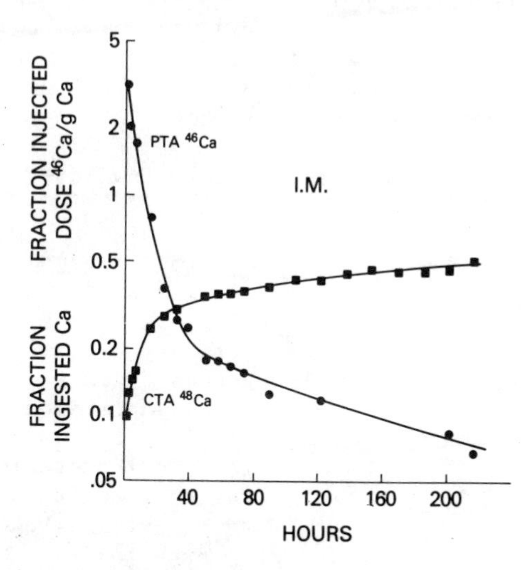

Figure 3. Time dependence of i.v. (diminishing) and oral isotope (increasing) for newborn subject.

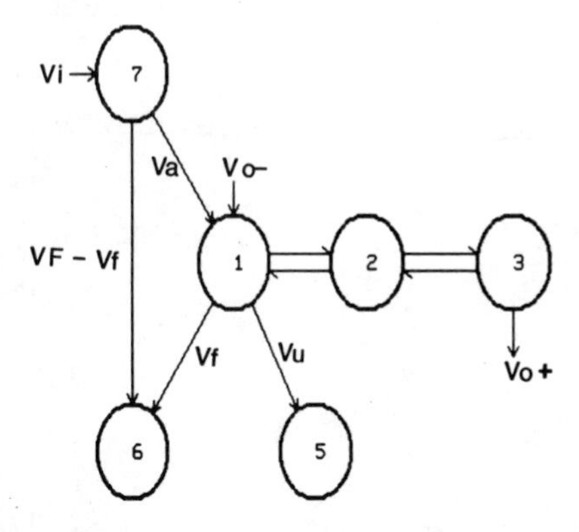

Figure 4. Three internal compartment serial model for calcium metabolism defining metabolic parameters.

Total calcium turnover is five to ten times greater in the newborn than in the adult. This is reflected in the fact that normalized calcium intake is 5-10 times higher in the newborn than in the adult, although the fraction absorbed (Va/Vi) is similar. A second observation is that in all five newborns, more calcium is lost through the endogenous fecal route than by urinary excretion. The reverse is true for typical adults. Finally, the is in positive balance for both bone and total organism calcium in contrast to the typical adult.

Table V. Symbols for Calcium Physiological Flow Rates

Symbol	Meaning (mg/kg/hr)
Vi	Ingestion
Va	Absorption
Vu	Urinary excretion
Vf	Endogenous fecal excretion
Vo+	Bone Accretion
Vo-	Bone resorption
VF	Total fecal excretion
Vbal	Total calcium balance =(Vi - Vu -Vf)
Vobal	Bone balance =(Vo+ - Vo-)

Table VI. Calcium Metabolism in Newborn and Adult

	Newborn (mg/kg/hr)	Adult (mg/kg/hr)
Vi	4.46	.575
Va	3.86	.249
Vu	0.077	.166
Vf	0.316	.101
Vo+	5.15	.323
Vo-	2.4	.344
VF	0.91	.432
Vbal	+3.47	-.023
Vobal	+2.75	-.021
Va/Vi	0.87	.43
Vf/Vu	4.0	.62

The power of this modeling technique might be better appreciated by considering a case of abnormal calcium metabolism, such as fibroplasia ossificans progressiva (FOP). The shape of the isotope dilution curves for this subject are notably different in the period of 1-2 days into the study as Figure 5 shows. They have a flattening in the first 20 hours that is not

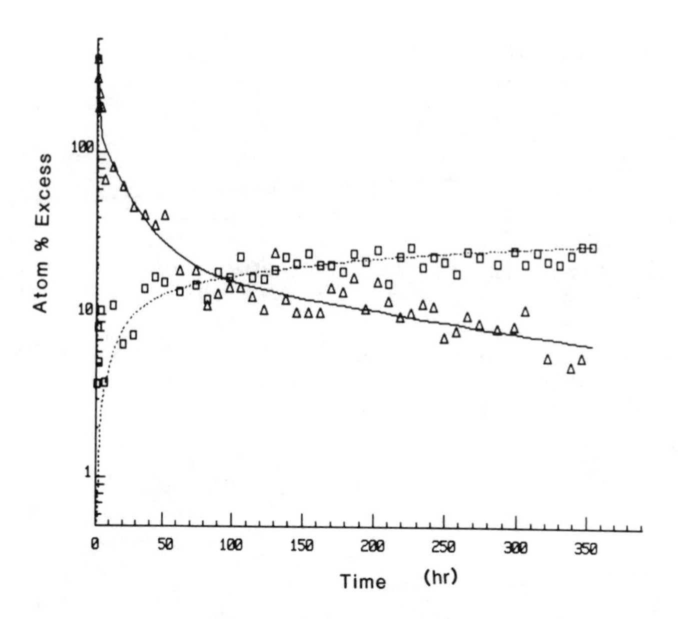

Figure 5. Time dependence of i.v. (diminishing) and oral isotope (increasing) for FOP subject.

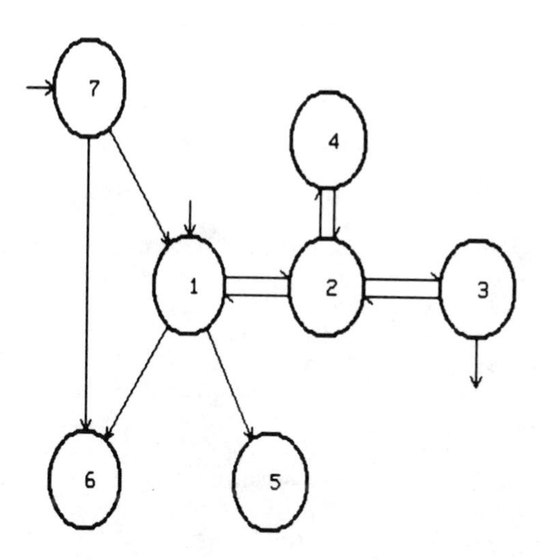

Figure 6. Four internal compartment serial model used to calculate calcium metabolic parameters for FOP subject.

seen in the normal tracer washout curves. The only way that a model could be made to fit the kinetic data for this subject was to include a fourth internal compartment of the sort shown in Figure 6. An additional compartment undergoing rapid exchange with another internal compartment, not contributing to the irreversible losses of those pools, is consistent with physiological intuition. The subject has greatly increased internal calcium stores that should be in contact with rapidly mixing calcium compartments, but these stores are undoubdtly at sites of very rapid growth which are almost certainly different from the slowly turning over bone pool represented by the loss from compartment 3 and the subsequent reentry into compartment 1. Data for this subject are given in Table VII along with the data for the normal subject used above.

Table VII. Calcium Metabolism in FOP and Adult

	FOP (mg/kg/hr)	Adult (mg/kg/hr)
Vi	.416	.575
Va	.169	.249
Vu	.009	.166
Vf	.312	.101
Vo+	.325	.323
Vo-	.478	.344
VF	.559	.432
Vbal	-.153	-.023
Vobal	-.153	-.021
Va/Vu	.41	.43
Vf/Vu	34.7	.62

Future

The studies of calcium metabolism in normal children are underway, with 15-20 children to be studied in the near future. The data generated from these studies will be used as the basis for studies of diseases of calcium metabolism in children. We anticipate being able to study a number of children with FOP and possibly contribute to innovative therapies for these children by monitoring changes in calcium kinetics. Improvements in the thermal ionization source for the quadrupole mass spectrometer are underway; a substantial improvement in sesitivity is expected from this effort. The model used for data analysis is under continuing investigation in order to improve its predictive and experimental design capabilities.

Literature Cited

1. A.L.Yergey, N.E.Vieira, J.W.Hansen, Anal. Chem., 52,1811 (1980).

2. D.L.Hachy, J.C.Blais, P.D.Klein, Anal.Chem., 52,1131 (1980).
3. J.S.Franklin, T.Clem, A.L.Yergey, in preparation.
4. L.J.Moore and L.A.Machlan, Anal.Chem., 44,2291 (1972).
5. Barnes, Murphy, Gramlich, Shields, Anal.Chem., 45,1881 (1973).
6. K.A.Kraus, F.Nelson, Proc. 1st Interntl. Conf. Peace. Uses Atomic Energy, 7,113 (1956).
7. A.Weintraub, A.L. Yergey, in preparation.
8. M.Berman and M.S.Weiss, SAAM Manual, USDHEW Publication.
9. R.Neer,M.Berman,L.Fisher and L.E.Rosenberg, J.Clin.Invest., 46,1364 (1967).
10. J.M.Phang, et al.,J.Clin.Invest., 48,67 (1969).
11. C.F.Ramberg, G.P.Mayer, D.S.Kronfeld and J.T.Potts, Am. J. Physio.,219, 1166 (1970).

RECEIVED January 31, 1984

3

Trace Element Utilization in Humans

Studied with Enriched Stable Isotopes and Thermal Ionization Mass Spectrometry

JUDITH R. TURNLUND

U.S. Department of Agriculture, Western Regional Research Center, Berkeley, CA 94710

Stable isotopes have recently proved to be valuable for determining absorption of zinc, copper, and iron in humans. Thermal ionization mass spectrometry, with its high degree of precision, has been successfully used for analysis of stable isotopes in samples from bioavailability studies. While precision of this analytical approach is excellent, analysis is time consuming. Isotopic ratios of iron can now be determined using an automated thermal ionization mass spectrometer, which markedly reduces analytical time. Two methods can be used to determine the amount of an isotopic spike in a sample with mass spectrometry: (1) by using the ratio of an enriched isotope to a natural isotope and the total mineral content of the sample, determined independently and (2) by using isotope dilution and determining two isotopic ratios. Several aspects of experimental design are critical when using stable isotopes to study mineral bioavailability. These include the level of isotope required to achieve an adequate enrichment in the tissue to be sampled, the effect of this level on mineral metabolism or absorption, complete intestinal transit time in studies using fecal monitoring, and sample homogeneity. Adaptation to type of diet or level of a nutrient, individual variability, and nutritional status must also be considered. Mineral absorption has been determined in several studies, comparing zinc, copper, and iron absorption from several types of diets and in different population groups. Results of these studies suggest a high level of phytate in the diet inhibits zinc, but not copper, absorption; pregnant women tend to absorb slightly, but not significantly more zinc and significantly more copper than nonpregnant women; and elderly men absorb less zinc, but not less copper or iron, than young men.

Several approaches have been used to determine absorption of trace elements in humans. The most frequently used method has been balance studies, in which the amount of a mineral ingested is compared with the amount eliminated in the feces. However, absorption calculated from total mineral eliminated in fecal collections generally differs greatly from true absorption, since some of the mineral eliminated in the feces is of endogenous origin (1). A number of other difficulties with metabolic balance studies, such as variation in intestinal transit time and inadequate analytical precision, limit their usefulness and often result in conflicting results (2).

Use of Isotopic Tracers

Many of the disadvantages of the balance approach to determining bioavailability can be eliminated or minimized by using radioactive tracers to measure absorption and utilization of trace elements. This approach has been used by a number of investigators to study zinc and iron absorption (3-7). Recently methods have been developed using enriched stable isotopes of trace elements to study absorption and utilization. A number of these studies, using a variety of methods and minerals, are described in this symposium. Using enriched stable isotopes provides many of the advantages of radioisotopic tracers without the exposure to radioactivity. Minerals contained in a specific meal or consumed on a specific day can be labeled and differentiated from minerals consumed at other times and from other endogenous minerals (8). Use of either stable or radioactive isotopes eliminates the problems of variablity of fecal flow, prolonged transit time, and excretion of endogenous minerals which confound results of absorption studies using the balance approach. The stable isotopes and several of the radioactive isotopes which have been used in human and animal studies are shown in Table 1. The natural abundance of stable isotopes and the half-lives of some of the most suitable radioisotopes are also included in Table I.

Both zinc and iron have several stable isotopes and radiosotopes well suited for use as tracers. Magnesium and copper have relatively abundant amounts of their stable isotopes. Therefore, higher levels of these isotopes are required to achieve adequate sample enrichment. Nevertheless, stable isotopes of these minerals are particularly attractive alternatives to radioisotopes, since the radioactive isotopes have very short half-lives. The half life of ^{28}Mg is only 21 hours. Half-lives of ^{64}Cu and ^{67}Cu are 12.9 and 61.9 hours respectively. Stable isotopes techniques cannot be used for manganese, since it has only one stable isotope.

Table I. Stable Isotopes

Mineral	Stable Isotope	Abundance %	Radio Isotope	Half-Life
Zinc	64	48.89	65	243.6 days (γ)
	66	27.81	$^{m}69$	13.9 hours (γ)
	67	4.11		
	68	18.57		
	70	0.62		
Iron	54	5.82	59	45.1 days (γ & $β^-$)
	56	91.66	55	2.6 years (X-Ray)
	57	2.19		
	58	0.33		
Magnesium	24	78.70	28	21 hours (γ & $β^-$)
	25	10.13		
	26	11.17		
Copper	63	69.09	64	12.9 hours ($β^-$)
	65	30.90	67	61.9 hours (γ & $β^-$)
Manganese	55	100	54	303 days (γ)

Selecting the Stable Isotope and Determing the Amount to Use

Ideally, stable isotopes selected for use as tracers are naturally low in abundance, such as ^{70}Zn and ^{58}Fe. These isotopes are also expensive. The cost of 1 mg of zinc containing 99% ^{70}Zn was $154 in 1983 (Oak Ridge National Laboratory), while 1 mg of zinc containg 99% ^{64}Zn cost only $0.55. However, the level of ^{64}Zn which occurs naturally, 48.89%, makes it impractical to satisfactorily enrich human samples with ^{64}Zn for absorption studies without greatly exceeding normal dietary levels of zinc. ^{67}Zn is another isotope that is useful for labeling. The price of ^{67}Zn is much less than ^{70}Zn, $5.25 for 1 mg of zinc containing 93% ^{67}Zn. Maximum precision of isotopic ratio determinations is achieved when the ratio between the two measured isotopes is between 1:10 and 10:1, and problems with background levels of the enrichment isotope are minimized when the abundance of the enriched isotope is at least doubled or 100% atom excess (9).

The amount and cost of ^{70}Zn and ^{67}Zn to meet these requirements for a sample expected to contain 100 mg of zinc can be calculated. When ^{68}Zn is used as the ratio isotope, 100 mg of zinc contains approximately 19 mg of ^{68}Zn. Less than 0.7 mg of ^{70}Zn would be required to double the ^{70}Zn content of the sample; however, 2 mg of ^{70}Zn would be required at a cost of $308, to achieve a ratio of ^{68}Zn:^{70}Zn between 1:10 and 10:1. The ^{68}Zn to ^{67}Zn ratio is naturally within the 1:10 to 10:1 range, and it remains in that range after doubling the natural ratio. Therefore, best precision can be achieved by doubling the natural ^{67}Zn, or using about 5 mg of ^{67}Zn at a cost of only $26.

To determine the amount of isotope required for an absorption study using fecal monitoring, the length of time of fecal collections must be considered. Polyethylene glycol (PEG) and radioisotopes were used for to determine complete intestinal transit time in the studies described in this paper. In one study PEG and radioisotopes were fed simultaneously and excretion patterns were similar for the two (1,9). Most individuals eliminate all PEG or unabsorbed radioisotopes within 12 days, but a few individuals have longer transit times. Therefore 12 days collection time can be used for calculations. The amount of mineral expected in the 12 day collection is used to determine the amount of isotope to feed for best analytical precision. A mineral such as copper cannot be fed at a level to meet ideal enrichment conditions without greatly exceeding normal dietary levels. This is because the least abundant isotope, ^{65}Cu, occurs naturally at the level of 30.9%. Because less than ideal enrichment of fecal collections is achieved with normal dietary levels, excellent precision of isotopic copper measurement is required for meaningful results. The precision of the absorption measurement can be improved by feeding ^{65}Cu for several days to increase the level of enrichment.

Determination of stable isotopes in other tissues requires additional considerations. To measure iron absorption via isotopic enrichment of erythrocytes requires labeling of 2.5 to 3.0 g of erythrocyte iron. To double the natural ratio of 100 mg of erythrocyte iron would require less then 0.4 mg of ^{58}Fe or less than 6 mg of ^{54}Fe. However, this does not achieve the ideal enrichment range for ^{58}Fe. About 9 mg of ^{58}Fe is required to achieve the ideal range. Since less than 10% of dietary iron is usually absorbed ten times the above amounts, approximately 60 mg of ^{54}Fe or 90 mg of ^{58}Fe have to be fed to optimally label 100 mg of erythrocyte iron. To label the 2.5 to 3.0 g of erythrocyte iron, extremely high levels of dietary iron would have to be used. For enrichment sufficient to measurement absorption with less then ideal conditions, but adequate to see differences in absorption, would require that iron isotopes be fed for several days. In this case, as with copper, a high degree of analytical precision is required.

Measuring stable isotopes in tissues such as blood serum or urine after feeding isotopes also requires the ability to measure less than ideal levels of enrichment. Therefore, excellent precision and accuracy are necessary. A study is in progress at the USDA Western Human Nutrition Research Center in San Francisco to establish the levels of isotopes that can be measured and followed in serum, urine and feces. Approximately 10% of circulating levels of zinc, copper, iron, calcium, and magnesium were injected into several young men in preparation for future mineral kinetic studies. Ideal enrichment levels may not be achieved when these amounts are injected, but injection of larger amounts could result in major changes in mineral metabolism (10).

Sample Collection and Preparation

Two critical aspects of fecal monitoring for absorption determinations are complete collection of unabsorbed isotopes and sample homogeneity. The times required for complete collections of unabsorbed isotopes vary markedly between subjects and also within the same subject. Therefore fecal makers are required to assure complete collections (9). The times required for complete elimination of a fecal marker have ranged from 6 to 18 days and included from 3 to 23 fecal samples in the experiments discussed later in this paper. Collection of only 80% of unabsorbed isotopes, particularly when absorption is low, will result in serious overestimation of absorption. In the case of iron, absorption of only 10% would appear to be nearly 30%, if 20% of the unabsorbed iron was not collected. A marker such as PEG, which can be measured quantitatively, must be used. Stools are collected until no trace of the inert marker can be detected. Recovery of less than 90% of the PEG would suggest that one or more collections were missed and absorption data would not be valid. Prior to sampling for analysis, the fecal pool must be homogenous in both isotopic distribution and in total mineral content. Both may vary considerably within the same stool and vary even more between stools. To assure sample homogeneity, the approach we have used in studies reported to date (1,9,11,12) has been to add deionized water to samples, homogenize thoroughly in a colloid mill, take subsamples while the colloid mill is running, and freeze for later analysis. However, after freezing and thawing of samples, some settling occurs, which makes it difficult to obtain a homogeneous sample. Settling does not affect isotopic homogeneity. It can alter mineral distribution in the sample, requiring numerous replicate determinations of total mineral content of samples. To avoid this problem, the procedure has been changed. The fecal pool is thoroughly mixed in a plastic container with added water using a polytron. Once the sample appears homogeneous, mixing is continued for 10 minutes. The fecal pools are then frozen, lyophilized, and crushed to a fine powder prior to sampling and preparation for analysis.

Calculation of Stable Isotope Content

Sample analysis by thermal ionization mass spectrometry (TIMS) results in measurement of isotopic ratios of minerals. Total mineral content of samples is then determined by one of two methods. One approach is to use flame atomic absorption spectrophotometry (AAS) to determine total mineral content of samples. Since AAS does not have the same level of precision as TIMS, a sufficient number of replicates is analyzed for a mineral content determination with a CV of within 1%. Alternatively, if a mineral has 3 or more isotopes and fractionation corrections are not made, the following procedure may be used. An individual is fed one isotope and another isotope is added to the sample prior to analysis to determine the total mineral content of the sample by dilution of the second isotope. In this way, both the amount of the isotope fed which is recovered in the feces and the total mineral content of the sample can be determined simultaneously. If fractionation corrections are to be made, a mineral must have at least four isotopes. Details of these procedures will be reported separately.

In general, formulas used to calculate the fraction of a stable isotope ingested which is recovered in feces using AAS to measure total mineral content are:

$$R_j^i = \frac{{}^{n}A_i \frac{{}^{n}M}{{}^{n}W} + {}^{s}A_i \frac{{}^{s}M}{{}^{s}W}}{{}^{n}A_j \frac{{}^{n}M}{{}^{n}W} + {}^{s}A_j \frac{{}^{s}M}{{}^{s}W}}$$

This is solved for $\frac{{}^{n}M}{{}^{s}M}$ by rearranging terms and simplifying.

Since ${}^{t}M = {}^{n}M + {}^{s}M$, and ${}^{t}M$ and $\frac{{}^{n}M}{{}^{s}M}$ are known, ${}^{s}M$ can be determined.

$${}^{s}M = \frac{{}^{t}M}{1 + \frac{{}^{n}M}{{}^{s}M}}$$

i, j and k = isotopes
R = measured ratio
A = isotopic abundance
M = mass of mineral
W = atomic weight of mineral
n = natural mineral
s = isotope spike fed to subjects
t = total mineral
d = isotopic diluent added to sample

The amount of the stable isotope spike which is recovered in the feces is subtracted from the total amount fed to determine the amount absorbed.

When isotope dilution is used to determine total mineral content, two ratios are measured. The following two equations are used to determine the two unknowns, (1) the total natural mineral content of the sample (^{n}M) and (2) the amount of isotope spike (^{s}M) recovered in feces:

$$R_j^i = \frac{{}^{n}A_i \frac{{}^{n}M}{{}^{n}W} + {}^{s}A_i \frac{{}^{s}M}{{}^{s}W} + {}^{d}A_i \frac{{}^{d}M}{{}^{d}W}}{{}^{n}A_j \frac{{}^{n}M}{{}^{n}W} + {}^{s}A_j \frac{{}^{s}M}{{}^{s}W} + {}^{d}A_j \frac{{}^{d}M}{{}^{d}W}}$$

and

$$R_j^k = \frac{{}^{n}A_k \frac{{}^{n}M}{{}^{n}W} + {}^{s}A_k \frac{{}^{s}M}{{}^{s}W} + {}^{d}A_k \frac{{}^{d}M}{{}^{d}W}}{{}^{n}A_j \frac{{}^{n}M}{{}^{n}W} + {}^{s}A_j \frac{{}^{s}M}{{}^{s}W} + {}^{d}A_j \frac{{}^{d}M}{{}^{d}W}}$$

The two equations are solved for the two unknowns, ^{s}M and ^{n}M.

When a mineral has more than two stable isotopes, experiments can be conducted using more than one isotope to label diets (13). Different isotopes can be used for different routes of administration, as has been done with radioisotopes of iron (14). Information can potentially be obtained on excretion of circulating mineral into the gastrointestinal tract by injecting stable isotopes, provided the injected dose is excreted similar to circulating mineral. Urinary excretion of minerals, and disappearance of the injected isotopes from the blood may also be measured. In combination with oral doses, these data can be used to advance the current knowledge of mineral metabolism and kinetics.

Thermal Ionization Mass Spectrometry Analysis

The method of isotope analysis I have selected is thermal

ionization, magnetic sector mass spectrometry (TIMS). This method was selected because the level of precision which can be achieved using TIMS is far better than precision which can be attained with other analytical methods (15). The improved precision has been demonstrated by measuring zinc isotopes in human plasma samples. The relative standard deviation of TIMS determinations of ^{70}Zn was 0.13%, while the relative standard deviation of neutron activation analysis (NAA) was 8.3% (16). The precision of TIMS was better than NAA by a factor of 64. In another human study, stable and radioisotopes of zinc and iron were fed to determine their absorption (8). Both TIMS and NAA were used for stable isotope analysis. The results of stable isotope analysis using TIMS were closer to the results using radioisotopes than those from NAA (8).

Two TIMS instruments have been used for the work described here. The instrument used most extensively is a single direction focusing magnetic sector mass spectrometer consisting of a sixty-degree magnetic sector with an ion radius of 30 cm (9). Recently iron determinations have been made using an automatic, computer controlled TIMS with a 13 sample turret. The iron content of a sample was determined with this instrument with a coefficient of variation of 0.6% using isotope dilution and several levels of isotope enrichment (17). Iron concentrations from TIMS were more precise, and were comparable to iron content determined using atomic absorption spectrophotometry, after atomic absorption values were corrected for recovery.

Experimental Diets

Absorption studies have been conducted using several dietary treatments. One dietary approach has been to use purified diets with egg albumen as the protein source. This approach reduces of day to day and between subject variability in the nutrient content of diets. It is well suited for use when levels of specific nutrients or dietary components must be carefully controlled and/or varied without changing any other factors in the diet. Since all other dietary components remain constant, any differences found between treatments can be ascribed to the altered nutrient under study.

Regular diets are used in some experiments, since they are most similar to diets consumed by a free-living population. However the problem of identifying a cause and effect relationship often limits their use in experiments designed to study specific nutrient effects. Alterations in the level of one nutrient in a regular food diet requires simultaneous changes in many nutrients and non-nutrient components of the diet. As a result, changes cannot be ascribed exclusively to the nutrient under study. For example, if zinc absorption from low zinc and a high zinc diet are to be compared using normal food diets, it may be possible to keep protein, carbohydrate and fat levels constant, but the food sources of these components will change. Mineral content of the two diets will differ widely, since foods low in zinc are likely to be low in other minerals as well. Levels of vitamins, phytate, fiber, and other components of foods may also vary. In addition, composition of regular foods can vary markedly, and constant nutrient intake may be compromised.

One approach which has been used to improve diet homogeneity, which is important for balance and absorption studies, is to use regular foods and then homogenized them in a blender. However, it is not known if absorption of minerals is the same when foods are homogenized as when they are not.

Factors Affecting Mineral Absorption

Another important consideration in designing mineral absorption experiments is the effect of adaptation to a specific type of diet or level of mineral. In mineral absorption studies conducted to date, increasing periods of adaptation to a diet or to a level of nutrient results in decreased variability between subjects. Absorption determined from a diet which is fed only at one meal or on one day may be very different from absorption studied after a period of adaptation to the diet is allowed.

Individual variability and nutritional status of individuals must also be considered when evaluating results of absorption studies. These effects are minimized when subjects are used as their own controls. Absorption studies of zinc, copper, and iron have been conducted with elderly men (9,11), young men, pregnant and nonpregnant women (1,12). Several types of diets have been used including purified diets with egg albumen as the protein source, homogenized food diets with protein from primarily animal or vegetable sources, regular food diets, and purified diets with α-cellulose or phytate added (18, 19).

Experimental Results

Trace mineral absorption by young and elderly men is shown in Table II. Zinc absorption was found to be significantly lower in elderly men than in young men who consumed a nearly identical diet. However, no difference in copper or iron absorption was seen between the young and elderly subjects.

When pregnant and nonpregnant women were fed diets containing protein from primarily animal or vegetable protein sources (Table III) pregnant women tended to absorb slightly, but not significantly more zinc and significantly more copper than nonpregnant women (1,12). Zinc absorption was similar from both diets, but fractional copper absorption was lower from the plant protein diet, which contained more copper than the animal protein diet.

Table II. Trace element absorption by young and elderly men

	Young Men	Elderly Men
	n = 6	n = 6
Zinc absorption (%)*	32.8 ± 1.9	18.0 ± 4.2**
Copper absorption (%)	26.5 ± 1.3	27.7 ± 0.31
Iron absorption (%)	8.3 ± 2.6	8.7 ± 3.6

* Mean ± SEM
** Significantly different ($P < 0.005$)

Table III. Zinc and copper absorption by pregnant and non-pregnant women.

	Nonpregnant women	Pregnant women
Zinc absorption (%)*		
Animal protein diet	23.8 ± 2.0 (5)**	24.1 ± 2.2 (5)
Plant protein diet	25.4 ± 1.7 (5)	26.6 ± 3.5 (4)
Copper absorption (%)		
Animal protein diet	41.2 ± 1.3 (5)	42.2 ± 2.5 (5)
Plant protein diet	33.8 ± 0.8 (5)	40.7 ± 3.0 (4)

* Mean ± SEM.
** Numeral in parentheses is n.

Zinc and copper absorption were compared in young men when α-cellulose or phytate were added to purified diets (Table IV). Copper absorption was not affected by the additions (19), but zinc absorption was significantly lower when phytate was included in the diet (18). Differences in copper absorption were between subjects also noted .

Table IV. Zinc and copper absorption from diets containing phytate and fiber.

	Basal* diet	Phytate added	α-cellulose added
Zinc absorption (%) (standard error of pooled mean = 3.54)	34.0	17.5**	33.8
Copper absorption (%) (standard error of pooled mean = 2.36	35.0	31.4	34.1

* n = 4 for each treatment
** Significantly lower (P < 0.02)

These studies demonstrated that differences due to population group or dietary treatment can be detected using stable isotopes with analysis by thermal ionization mass spectrometry. Further research using stable isotopes and TIMS promises to provide valuable information on mineral absorption and utilization.

Literature Cited

1. Turnlund, J. R.; Swanson, C. A.; King, J. C. J. Nutr. 1983, In press.
2. Kelsay, J. L. Cereal Chem. 1981, 58, 2-5.
3. Spencer, H.; Rosoff, B.; Lewin, I.; Samachson, J. in: "Zinc Metabolism"; Prasad, A., Ed.; Charles Thomas: Springfield, 1966, pp 339-362.
4. Aamodt, R. L.; Rumble, W. F.; Johnson, G. S.; Markley, E. J.; Henkin, R. I. Am. J. Clin. Nutr. 1981, 34, 2648-2652.
5. Sandstrom, B.; Cederbland, A. J. Nutr. 1980, 33, 1778-1783.
6. Cook, J. D.; Monsen, E. L. Am. J. Clin. Nutr. 1976, 29, 859-867.
7. Rossander, L.; Hallberg, L.; Bjorn-Rasmussen, E., Am. J. Clin. Nutr. 1979, 32, 2484-2489.
8. Turnlund, J. R. Sci. Total Environ. 1983, 28, 385-392.
9. Turnlund, J. R.; Michel, M.C.; Keyes, W. R.; King, J. C.; Margen, S. Am. J. Clin. Nutr. 1982, 35, 385-392.
10. Klein, P.D.; Personal communication.
11. Turnlund, J. R.; Michel, M. C.; Schutz, Y; Margen, S. Am. J. Clin. Nutr. 1982, 36, 587-591.
12. Swanson, C. A.; Turnlund, J. R., King , J. C. J Nutr. 1983, In Press.

13. Janghorbani, M.; Ting, B. T. G.; Young, V. R. Br. J. Nutr. 1981, 46, 395-402.
14. Hallberg, L.; Bjorn-Rasmussen, E.; Garby, L.; Pleehachinda, R., Suwanik, R. Am. J. Clin Nutr. 1978, 31, 1403-1408.
15. Turnlund, J. R.; King, J. C. in "Nutritional Bioavailability of Zinc"; Inglett, G. E., Ed.; ACS SYMPOSIUM SERIES No. 210, American Chemical Society; Washington, D.C., 1983; pp. 31-40.
16. Janghorbani, M; Young, V. R.; Gramlich, J. W., Machlan; L. A. Clin. Chem. Acta 1981, 114, 163-171.
17. Turnlund, J. R., Gong, B.; Reager, R. D. Pacific Conf. Chem. Spect. 1982, No. 181.
18. Turnlund, J. R.; King, J. C.; Fed. Proc. 1983, 42, 822.
19. Turnlund, J. R., King, J. C.; Am. J. Clin. Nutr. 1983, 32, 716 and Clin. Res. 1983, 31, 624a.

RECEIVED January 20, 1984

4

Mössbauer Spectroscopy in Nutritional Research

LEOPOLD MAY

Department of Chemistry, Catholic University of America, Washington, DC 20017

Mössbauer spectroscopy provides a probe of atoms that is sensitive to changes in oxidation and spin state and the configuration of the ligands around the atom. It is a nondestructive method and specific for the isotope being examined. The number of elements that can be examined by Mössbauer spectroscopy is limited but includes several that are important in nutritional studies, such as cobalt and iron. The basic principles and techniques of Mössbauer spectroscopy will be described, including the isotope enrichment procedures that have been used. Its use in nutritional research will be illustrated with structural studies of Vitamin B_{12} using cobalt emission Mössbauer spectroscopy and the identification of the iron compounds in wheat bran using iron absorption Mössbauer spectroscopy.

Mössbauer or nuclear gamma resonance fluorescence spectroscopy, akin to other forms of spectroscopy, is the study of the radiation emitted or absorbed in the transition between two or more energy levels (1). These levels are those of the nucleus, and the energy of radiation is found in the gamma ray region. In the observation of these changes, the energy of the transition in the source nucleus (radiation emitter) and the energy of the transition in the absorber nucleus must be equal (Figure 1). Since the emitting or absorbing ray imparts recoil energy to the nuclei causing a decrease in the energy of the gamma ray, it is difficult to observe resonant absorption. Mössbauer (2) found that in crystalline lattices at temperatures significantly below the Debye temperature of the crystal, the nucleus could emit or absorb gamma radiation with little or no recoil energy, and resonance could be observed. If the nucleus is placed in a

0097-6156/84/0258-0053$06.00/0

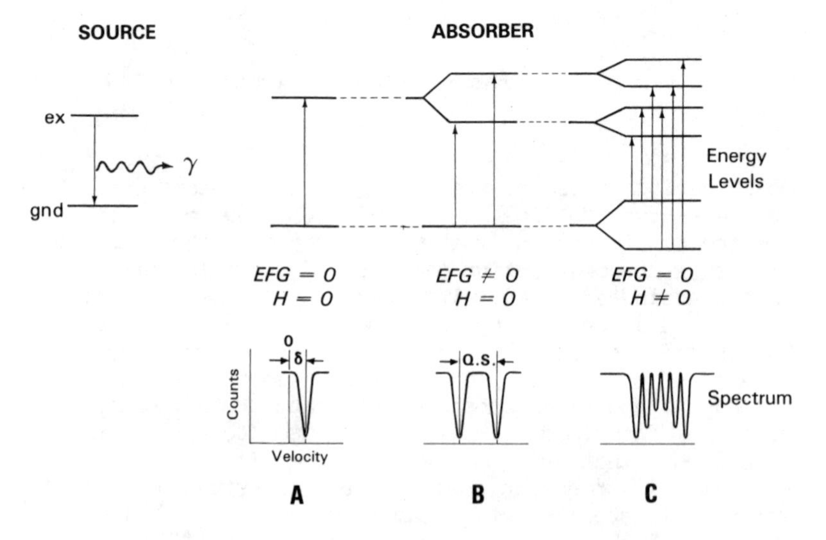

Figure 1. Nuclear transitions and Mossbauer spectra. EFG, electric field gradient; H, magnetic field; δ, isomer shift; and Q.S., quadrupole splitting.

different electronic (chemical) environment, as the absorber in Figure 1, the transition energy of the absorber is changed destroying resonance. To restore resonance, if the source and absorber do not have the identical transition energies, i.e., chemical environment, additional energy must be added to or subtracted from the gamma rays of the source, as in Figure 1. The required energy can be supplied by moving the source and absorber relative to each other, which adds Doppler motion energy to the system permitting the resonance condition to be established.

If the nucleus has a quadrupole moment and its electric field gradient (EFG) about the nucleus is asymmetric, the nuclear energy levels are split as shown for iron in Figure 1B. This gives rise to a doublet in the Mössbauer spectrum. A Zeeman effect can also be observed in a magnetic field (external or internal) giving rise to a six-line spectrum for iron as illustrated in Figure 1C. The simultaneous presence of both an EFG and magnetic field with the nucleus causes the resulting spectrum to be asymmetric.

If the source and the absorber are identical, resonance absorption will occur with both stationary. However, if they are not identical (Figure 1A) resonance absorption will appear at a Doppler velocity different from zero. This difference is the isomer shift, δ, and is related to the s-electron density at the nucleus. If the spectrum is split, as in Figure 1B, then the difference between the velocities of the two peaks is the quadrupole splitting (Q.S.). This splitting is influenced by the configuration of the electronic environment around the nucleus, and its magnitude yields information concerning the bonding of the atom. The amount of the splitting of the peaks in a magnetically split spectra, as in Figure 1C, can be related to the internal magnetic field in paramagnetic and ferromagnetic substances.

Essential Nutritional Elements and Mössbauer Nuclides

The Mössbauer effect has been observed with about 50 elements, including several that are accepted as essential for nutrition (3), shown as shaded in Figure 2. Those that in principle may be observed in the absorption mode are shown with small dots and those that in principle may be observed in the emission mode are shown with large dots. The elements that have been observed in either the absorption or the emission mode include cobalt, iodine, iron, and zinc. The cobalt Mössbauer spectrum has been observed in the emission mode, and the spectra of the other elements have been observed in the absorption mode.

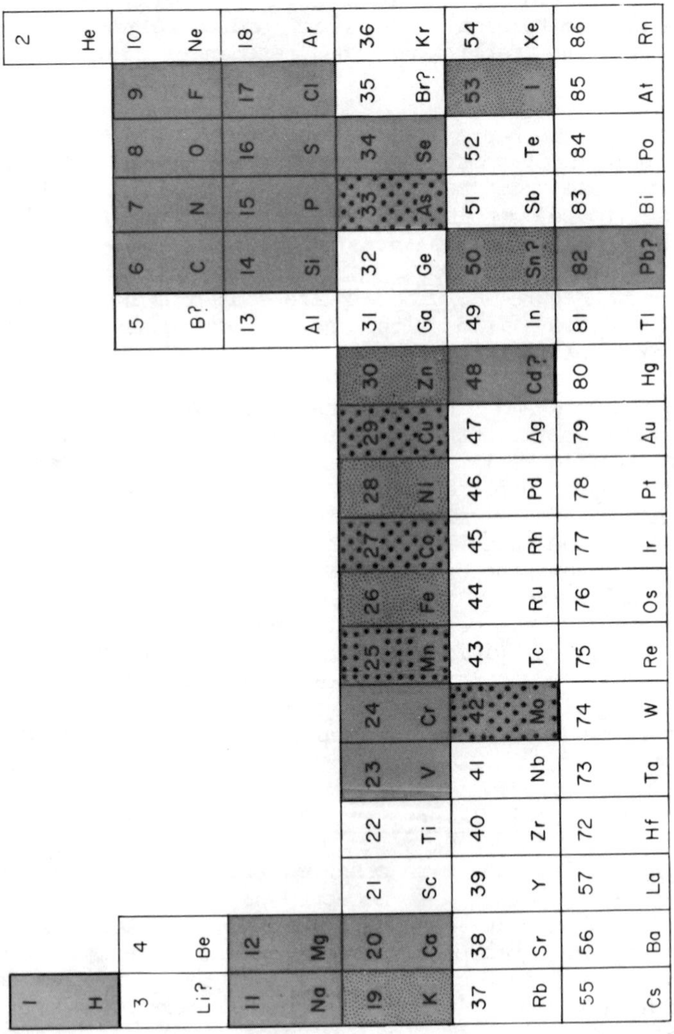

1 H																	2 He
3 Li?	4 Be											5 B?	6 C	7 N	8 O	9 F	10 Ne
11 Na	12 Mg											13 Al	14 Si	15 P	16 S	17 Cl	18 Ar
19 K	20 Ca	21 Sc	22 Ti	23 V	24 Cr	25 Mn	26 Fe	27 Co	28 Ni	29 Cu	30 Zn	31 Ga	32 Ge	33 As	34 Se	35 Br?	36 Kr
37 Rb	38 Sr	39 Y	40 Zr	41 Nb	42 Mo	43 Tc	44 Ru	45 Rh	46 Pd	47 Ag	48 Cd?	49 In	50 Sn?	51 Sb	52 Te	53 I	54 Xe
55 Cs	56 Ba	57 La	72 Hf	73 Ta	74 W	75 Re	76 Os	77 Ir	78 Pt	79 Au	80 Hg	81 Tl	82 Pb?	83 Bi	84 Po	85 At	86 Rn

Figure 2. Periodic table. Shaded, essential elements; small dots, may be observed in Mossbauer absorption mode; and large dots, may be observed in Mossbauer emission mode.

Requirement for Enrichment

Many of the Mössbauer nuclides occur in nature, for example, ^{127}I. With other elements, the percentage of the Mössbauer nuclide is low (2.17% for ^{57}Fe and 4.11% for ^{67}Zn). In the case of cobalt, the radioactive nuclide, ^{57}Co, must be inserted into the material being examined. It is desired to have the amount of the Mössbauer nuclide in the sample large so that the Mössbauer effect will be large. In many biological samples, the Mössbauer effect is small, so it is advantageous to have a large concentration of the Mössbauer nuclide. This requires that the concentration of the nuclide in the sample be increased. The enrichment can be accomplished by removing the metal and replacing it with the Mössbauer nuclide. This can be done directly if the compound does not denature when the metal is removed. For cobalt emission spectroscopy, the cobalt in the sample must be replaced with ^{57}Co, the Mössbauer nuclide. If the compound is to be separated from an organism, for example, plant or bacteria, the organism can be grown in a medium containing a high concentration of the Mössbauer nuclide. If a compound is to be enriched in an animal, the feed can be enriched with the Mössbauer nuclide, or the appropiate enriched compound can be injected into the bloodstream.

Application to Vitamin B_{12}

One of the important applications of Mössbauer spectroscopy is in the determination of bonding of the Mössbauer nuclide. Since the cobalt Mössbauer spectrum can only be observed in the emission mode, the radioactive ^{57}Co isotope must be inserted into the compound. The compound then serves as the source with the absorber being an iron compound such as potassium ferrocyanide. This requires the removal of the normally occurring cobalt isotope and replacement with the Mössbauer nuclide without altering the structure of the compound. For example, Vitamin B_{12} (cyanocobalamin) that contains cobalt bound to a corrin ring system with cyanide ligand in the sixth position was examined. The electron capture decay of a ^{57}Co atom with the emission of the gamma rays triggers events in which energy is deposited into the molecule resulting possibly in fragmentation of the molecule. Nath, et al. (4) observed that this did not occur with the cobalamins. The oxidation state and binding of the daughter atom, ^{57}Fe, remains the same as the ^{57}Co. Thus, the oxidation state and binding of the cobalt atom has been studied in Vitamin B_{12} and derivatives with the cyanide ligand replaced by other ligands such as aquo, acetato, etc. (4-6). Emission Mössbauer spectroscopy of some of these derivatives has been used to study ethanolamine ammonia lyase, a coenzyme B_{12}-dependent enzyme (7).

Identification of Endogenous Iron in Wheat Bran

Another application of Mössbauer spectroscopy is illustrated by the identification of the endogenous iron in wheat bran (8). Morris and Ellis (9) found that more than 60% of the iron in the bran was extracted in association with phytic acid as monoferric phytate. The bioavailability of the isolated iron product and the synthetic phytate did not differ (9,10). Knowledge of the chemical nature *in situ* of the iron in the bran might aid in explaining the action of bran on iron absorption. The Mössbauer spectra of wheat seeds, wheat bran, monoferric phytate, and diferric phytate were measured. To enrich the iron in the wheat seeds, the wheat was grown in a medium containing ^{57}Fe. The spectra are shown in Figure 3, and the results in Table I. It can be seen that the Mössbauer parameters of the spectra of seeds and bran agree within experimental error with the parameters found in the spectrum of monoferric phytate. These results indicate that the iron associated with the phytate *in situ* is in the same combination as in monoferric phytate.

Table I. Mössbauer Spectral Parameters of Wheat Seeds, Wheat bran, and Ferric Phytates*

Sample	Quadrupole Splitting	Isomer Shift
Wheat seeds	0.55	0.76
Wheat bran	0.56	0.77
Monoferric phytate	0.55	0.77
Diferric phytate	0.60	0.76

*Parameters are in mm/s relative to sodium nitroprusside and were measured at 80 K (8).

Although the number of Mössbauer nuclides for the study of the nutritionally essential elements is limited, these examples illustrate the information and use that Mössbauer spectroscopy can provide in nutritional research.

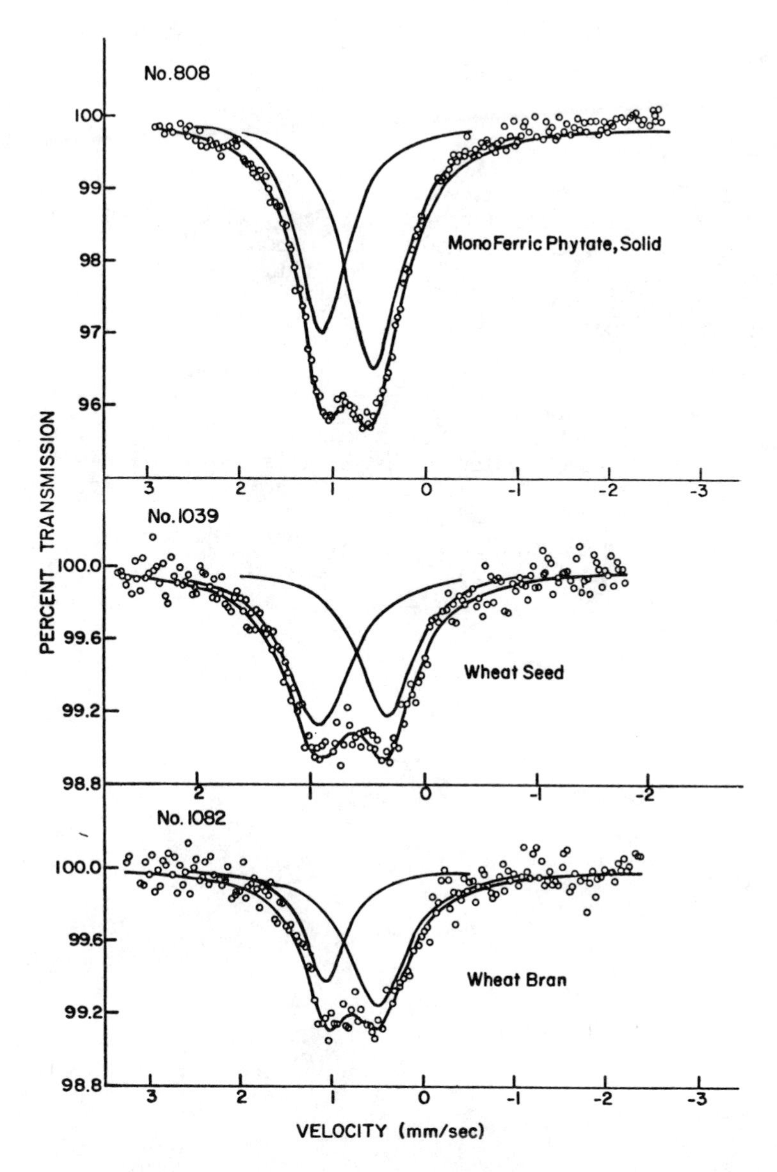

Figure 3. Mossbauer spectra of monoferric phytate (top), wheat seed (middle), and wheat bran (bottom). (Reproduced from Ref. 8. Copyright 1980, American Chemical Society.)

Acknowledgment

I thank Dr. Eugene R. Morris for his suggestions and assistance.

Literature Cited

1. May, L. "An Introduction to Mössbauer Spectroscopy"; Plenum: New York, 1971; Greenwood, N.N.; Gibb, T. C. "Mössbauer Spectroscopy"; Chapman and Hall: London, 1971.
2. Mössbauer, R. L.; Z. Physik., 1958, 151, 124-143.
3. Mertz, W.; Science 1981, 213, 1332-1338.
4. Nath, A.; Harpold, M.; Klein, M. P.; Kündig, W.; Chem. Phys. Lett. 1968, 2, 471-476.
5. Mullen, T.; Mössbauer Eff. Method. 1970, 5, 95-105.
6. Inoue, K.;Nath, A.; Bioinorg. Chem. 1977, 7, 159-167.
7. Cardin, D. J.; Joblin, K. N.; Johnson, A. W.; Lang, G.; Lappert, M. F.; Biochim. Biophys. Acta 1974, 371, 44-51.
8. May, L.; Morris, E. R.; Ellis, R.; J. Agri. Food Chem. 1980, 28, 1004-1006.
9. Morris, E. R.; Ellis, R.; J. Nutri. 1976, 106, 753-760.
10. Lipschitz, D. A.; Simpson, K. M.; Cook, J. D.; Morris, E. R.; J. Nutri. 1979, 109, 1154-1160.

RECEIVED January 20, 1984

5

Intrinsic Labeling of Edible Plants with Stable Isotopes

CONNIE M. WEAVER

Department of Foods and Nutrition, Purdue University, West Lafayette, IN 47907

Nutritionists and food scientists are seeking reliable and efficient methods to study bioavailability of trace elements from various foods. Among the several methods available, the most promising for human use involves labeling of foods with stable isotopes. This approach allows safe testing in human population groups and use of foods prepared in normal food handling and processing facilities. However, because of the high cost of stable isotopes, efficient labeling of foods is imperative. This paper will describe a non-cycling hydroponic system we have developed to incorporate isotopes into edible portions of plants. This system has been successfully employed in the efficient labeling of soybeans and wheat with isotopes of zinc, iron, and selenium.

Bioavailability of minerals from foods is most directly assessed by following the fate of the mineral after consumption of foods endogenously labeled with the mineral of interest. Labeling foods with stable isotopes rather than radioisotopes offers the advantage of using humans rather than animal models without the risks associated with radiation exposure. Also, foods can be prepared in normal food handling and processing facilities with use of stable isotopes. Therefore, the use of stable isotope tracer techniques allows bioavailability of trace elements to be studied safely in all human population groups using foods processed under normal conditions and in the form in which it is normally consumed.

Intrinsic vs Extrinsic Labeling Techniques

Intrinsic labeling is time consuming and specialized. Extrinsic labeling whereby a mineral isotope is mixed with a food prior to consumption is an appealing alternative to intrinsic labeling. The extrinsic labeling approach assumes that the extrinsic label

0097-6156/84/0258-0061$06.00/0

completely exchanges with the mineral endogenous to the food. The validity of this assumption requires testing against intrinsically labeled foods for each mineral and testing condition.

Extrinsic and intrinsic non-heme radioiron has been shown to exchange for a number of foods when the extrinsic label is thoroughly mixed with the food being tested prior to consumption (1,2). Using the rat model, an extrinsic labeling approach gave a similar assessment as an intrinsic labeling approach of zinc bioavailability for a few foods (3,4). However, when $^{65}ZnCl_2$ was mixed with a soy flour based diet rather than with the soy flour prior to incorporation into the diet, retention of ^{65}Zn was significantly different between the intrinsic and extrinsic labels (4). Thus, the potential of the extrinsic label for giving an accurate assessment of bioavailability from a specific food seems dependent on the manner of incorporation of the extrinsic label.

In humans, zinc from chicken intrinsically labeled with the stable isotope ^{68}Zn was absorbed to a greater extent than was an extrinsic tag, $^{70}ZnCl_2$ (5). Perhaps extrinsic and intrinsic labels do exchange in many instances where the labeled fraction is a small portion of the total mineral content such as in the case for radiosotopes. However, stable isotope labeling involves replacing a much larger fraction (mg quantities) of the total mineral with the label. These larger pools may not entirely exchange. Further testing of the validity of extrinsic labeling approach in stable isotope methodology is required for all minerals.

Intrinsic Labeling Methodology

Growing Conditions. Foods must be highly enriched with stable isotopes to be useful as labels for bioavailability experiments in order to accurately distinguish the label from naturally occuring isotopes. However, stable isotopes are expensive and their supply limited. Therefore, the labeling methodology must be as efficient as possible.

Labeling of plants via a nutrient solution is more efficient than by soil application due to the possible adherance of the added mineral on soil components. Stem injection may be an efficient labeling technique, but until the mineral complexes which are transported in the vascular system are clearly defined, this technique may not result in deposition of isotopes which are characteristic of field grown crops.

For the purpose of enriching plants with stable isotopes, it was necessary to develop a hydroponic technique which requires a small amount of nutrient solution per plant and which maintains the same nutrient solution from the addition of the stable isotope until maturity of the plant. The expense of stable isotopes prohibits using a great volume of nutrient solution or discarding nutrient solution before maximal uptake of the isotope is achieved.

The method I have developed for enriching with stable isotopes is a pot culture technique. Seeds are germinated in Perlite and seedlings are transferred to plastic pots (14 cm diameter and 19 cm depth) containing two liters of a modified Hoagland-Arnon (6) nutrient solution. The outer surfaces of the pots are sprayed with black paint or covered with black plastic to inhibit growth of algae. Four soybean or eight wheat seedlings per container are placed in holes in the lids and supported with foam rubber. During the entire growth cycle, solutions are aerated and deionized water is used to replace losses due to transpiration and evaporation. The pH is maintained between 5.8 and 6.0 with 1 N HCl or 1 N KOH. Conductivity is maintained with complete nutrient solution as needed; usually this requires replentishing with nutrient solution twice weekly. Nitrate, potassium, and phosphorus levels are monitored and adjusted as needed, with 1 M $Ca(NO_3)_2$, 1 M KNO_3, and 1 N KH_2PO_4. This system has been employed in a variety of greenhouse conditions and outside during the summer. For the experiments described here, plants were grown in the greenhouse where day temperatures were approximately 27°C. Supplemental lighting supplied 146 microeinsteins/M^2 (cloudy day) with 14 hour light period. Once stable isotopes are added, solutions are not replaced.

To determine the optimal conditions for efficiently labeling plants with stable isotopes, several experiments employing radioisotopes and the non-cycling hydroponic system were designed.

When to Dose. To determine whether efficiency of incorporation of an isotope is dependent on time of tracer application, 1.1 x 10^7 cpm/pot radiozinc ($^{65}ZnCl_2$), radioselenium ($^{75}SeO_3^{--}$), or radioiron ($^{59}FeCl_3$) were administered with a stable carrier to provide 1.5 ppm Zn, 0.5 ppm Se, or 0.5 ppm Fe either 5 weeks from germination or during flowering or fertilization of soybeans or wheat. The radionuclides were not chelated because under the same conditions employed here chelation did not improve uptake of ^{59}Fe by soybeans from the nutrient solution (7). The distribution of the radioactivity among the plant parts and the efficiency of incorporation of the dose into the edible tissues as a function of time of application are given in Tables I and II.

The relative order of efficiency of accumulation of the three radionuclides into edible tissues was Zn>Fe>Se. Most of the ^{65}Zn transported to the above ground plant parts was accumulated by the seeds, however, much of the dose of each of the radionuclides remained in the roots and nutrient solution. The calculated percent of the applied dose taken up by soybean leaves is underestimated since some of the leaves abscised as the plants senesced and they were discarded. Zinc and iron have been classified as partially mobile in the phloem (8). Translocation of iron within plants is poor since new growth of plants require a continuous supply of iron via the xylem or from external applications (9). Zinc deficiencies of plants can be corrected by applying $ZnSO_4$ in a dilute spray. However, iron deficiences of plants are usually difficult to correct which implies a lesser mobility within the plant.

Table I. Concentration and Percent of Applied Dose of ^{65}Zn, ^{75}Se and ^{59}Fe Accumulated by Plant Part in Soybeans

	^{65}Zn Added during			
	Vegetative period		Flowering	
Plant Part	cpm/g[1] $\bar{X} \pm S$	% dose in plant part[2]	cpm/g $\bar{X} \pm S$	% dose in plant part[2]
Cotyledons (seeds minus hull)	36,720±2,090[a]	20.1	55,842± 6,636[a]	25.3
Hull	34,345±4,070[a]	1.2	63,439±24,735[a]	2.3
Pods	5,850± 566[b]	1.4	12,772± 2,320[a]	2.8
Leaves	35,468±7,692[a]	3.7	50,835±10,425[a]	2.1
Stem	4,636±1,006[b]	1.6	9,621± 1,720[a]	1.9
Roots[3]	7,399± 399	1.4	11,758± 676	1.5

Table I. Continued

Plant Part	^{75}Se Added during Vegetative Period		Flowering		^{59}Fe Added during Vegetative Period		Flowering	
	cpm/g $\bar{X} \pm S$	% dose in plant part[2]	cpm/g[1] $\bar{X} \pm S$	% dose in plant part[2]	cpm/g[1] $\bar{X} \pm S$	% dose in plant part[2]	cpm/g[1] $\bar{X} \pm S$	% dose in plant part[2]
Cotyledons (seeds minus hull)	25,412±1103[a]	13.4	15,834±3,383[a]	4.6	25,413± 1,807[a]	13.9	39,871	8.3
Hull	14,768± 648[b]	0.48	8,310±1,938[a]	0.16	117,981±12,089[b]	6.3	359,088	4.1
Pods	5,044± 316[c]	1.0	3,617±1,163[a]	0.5	9,063± 1,408[a]	1.3	6,658±2,659	1.0
Leaves	8,288± 910[c]	2.3	6,705±2,396[a]	2.7	10,014± 1,733[a]	4.3	19,259± 668	6.3
Stem	7,401± 886[c]	2.6	5,788±1,157[a]	1.8	10,595± 703[a]	4.2	14,118±4,554	4.7
Roots[3]	131,558±7688	29.9	298,978±2,514	43.0	86,855± 4,630	17.2	194,192±4,630	22.7

[1] Mean and standard deviation of samples from each plant in the container on a dry weight basis; different superscripts within colums denote significant differences ($P<0.01$)

[2] The percent dose accumulated by that plant part totaled for all 4 of the plants in that container.

[3] Roots were harvested as one unit from each treatment container and the mean concentrations of three aliqots per treatment were not compared to other plant tissue concentrations by analysis of variance.

Table II. Concentration and Percent of Applied Dose of ^{65}Zn and ^{75}Se Accumulated by Plant Part in Wheat

Plant Part	^{65}Zn added during: Vegetative period		Fertilization	
	cpm/g[1] $\bar{X} \pm S$	% dose in plant part[2]	cpm/g[1] $\bar{X} \pm S$	% dose in plant part[2]
Grain	111,677±30,172[a]	14.1	353,425±21,034[a]	25.6
Spike minus grain	39,309± 5,109[b]	1.9	229,949±13,273[b]	6.7
Leaves	79,962± 6,880[a]	7.3	222,973±13,350[b]	12.3
Stems	20,960± 1,847[b]	1.9	110,142± 7,745[c]	6.4
Roots[3]	124,753±54,181	7.5	381,890±37,921	15.4

Table II. Continued

Plant Part	^{75}Se added during Vegetative period		Fertilization	
	cpm/g[1] $\bar{X} \pm S$	% dose in plant part[2]	cpm/g[1] $\bar{X} \pm S$	% dose in plant part[2]
Grain	100,594±62,992[a]	3.1	49,274±3,256[a]	6.3
Spike minus grain	50,821± 1,686[a]	1.4	23,322±2,271[b]	1.2
Leaves	67,628± 2,709[a]	5.0	33,891±7,342[a,b]	2.2
Stems	44,068± 2,503[a]	5.3	21,081±1,366[b]	1.7
Roots[3]			455,226±37,853	21.3

[1]Mean and standard deviation of samples from each plant in the container on a dry weight basis; different superscripts within columns denote significant differences (P 0.01).

[2]The percent dose accumulated by that plant part totaled for all 8 of the plants in that container.

[3]Roots were harvested as one unit from each treatment container and the mean concentrations of three aliquots per treatment were not compared to other plant tissue concentrations by analysis of variance.

Little is known about the mobility of selenium in plants. The lesser uptake of selenium may be due to a lower mobility of selenium, to the formation of an insoluble complex between selenite and iron at the acidic pH of the nutrient solution or to selenium toxicity. The large amount of ^{75}Se accumulated by the roots of soybeans may reflect a selenium toxicity.

A greater percent of the ^{65}Zn dose was accumulated by the seeds of soybeans and wheat when the radionuclide was administered during the reproductive period of the plant. More ^{75}Se was accumulated by edible portions of wheat when administered during fertilization, but the reverse was true for soybeans. ^{59}Fe was also translocated to soybean seeds more efficiently when administered prior to flowering. When seeds are forming, they serve as a nutrient sink and nutrients are readily deposited in the reproductive tissues. Since iron deposited in leaves while plants are still growing is supposedly relatively immobile for translocation to the grain, one would expect a greater accumulation by seeds when applied during flowering. Perhaps iron at the roots was competing with the ^{59}Fe for the uptake sites and saturation of the sites increased with age of the plant.

Yields were not affected by any of the treatments except for ^{75}Se in wheat (Table III). The $Na^{75}SeO_3$ was administered with

Table III. Weights by Plant Parts of Wheat Exposed to ^{75}Se[1]

Plant Part	Controls (unlabeled)	^{75}Se added during Vegetative Period	^{75}Se added during Fertilization
	Dry wt (g) $\bar{x} \pm s$	Dry wt (g) $\bar{x} \pm s$	Dry wt (g) $\bar{x} \pm s$
Grain	2.053 ± 0.603^a	0.666 ± 0.427^b	$3.544 \pm 1/593^a$
Spike minus grain	0.527 ± 0.188^a	0.382 ± 0.206^a	1.395 ± 0.540^b
Leaves	1.083 ± 0.378^a	1.026 ± 0.308^a	1.812 ± 0.328^b
Stem	1.818 ± 0.792^a	1.068 ± 0.313^b	2.278 ± 0.373^a

[1]Means and standard deviations for 8 plants in that container; different superscripts within rows denote significant differences ($P<0.05$) according to Duncans Multiple Range Test.

sufficient $NaSeO_3$ to provide 0.5 ppm Se. The low yields of wheat exposed to 0.5 ppm Se during the vegetative period may be responsible for the low efficiency of incorporation of ^{75}Se with the grain.

To further define the stage of development at which optimal efficiency of labeling edible portions of soybeans and wheat, a series of pots were dosed sequentially beginning with the first appearance of flowering and ending with beginning of senescence (Figures 1-5). Stable carriers were not added with the single dose of radionuclides (1.1 x 10^7 cpm/pot) beyond what was present in the complete nutrient solution: 0.05 ppm Zn, 2.5 ppm Fe and no added Se. Seeds from the 4 soybean plants in each pot or 8 wheat plants in each pot per treatment were pooled and assayed for radioactivity on a Beckman 4000 counting system with counting efficiencies of 27, 48, and 33% for ^{65}Zn, ^{59}Fe, and ^{75}Se, respectively. As in the previous experiment, more ^{65}Zn was accumulated by seeds of wheat (Figure 2) and soybeans (Figure 4) than ^{75}Se (Figures 1 and 3) or ^{59}Fe (Figure 5).

No overall conclusion can be made that would allow a recommendation for dosing at a particular stage of development for all radionuclides and plants studied. Delaying administration of the label until senescence of soybeans was apparent (leaves began turning yellow at 8 weeks and pods at 9 weeks after flowering) resulted in a lower labeling efficiency for ^{75}Se (Figure 3), ^{65}Zn (Figure 4), and ^{59}Fe (Figure 5) so that labeling should begin prior to this period for maximum incorporation into seeds. It is tempting to recommend specific periods of application for some of the radionuclides ie.e. ^{65}Zn 20 days from anthesis for wheat (Figure 2) and 6 weeks from flowering for soybeans (Figure 4) or to recommend avoiding dosing during certain stages of development i.e. weeks 4 and 5 from flowering of soybeans for application of ^{75}Se (Figure 3) and ^{59}Fe (Figure 5) respectively. However, these data were collected on single pots at each dosing period for each treatment and were intended to use to establish trends rather than for literal use of the percentage dose accumulated by the seeds. The variety of plant, growing season, day length, temperature, etc. would affect the optimal time of uptake. Biological variation in uptake of minerals from plant to plant even in the same pot is great as evidenced by Tables I and II.

Representative wheat spikes or soybean flowers from each pot in this experiment were tagged with a paper label dated with their flowering date. Representative fruits were also identified with tags marked with colored tape which served as a code for the stage of development of the seeds at the time of dosing. These seeds were kept separate when harvested at maturity for assessment of radioactivity. The concentration of the radionuclides in seeds that were at different stages of development when dosing occurred for soybeans and wheat is given in Tables IV and V. If the concentration of radioactivity in the mature seeds was higher for any radionuclide at any particular stage of fruiting, efficiency of incorporation of a label should be higher when most of the fruits were at that stage of development. No stage of seed maturity can be selected as the optimal dosing time. The data selected for Tables IV and V were from pots with the widest range of stage of seed development at time of dosing. Only data

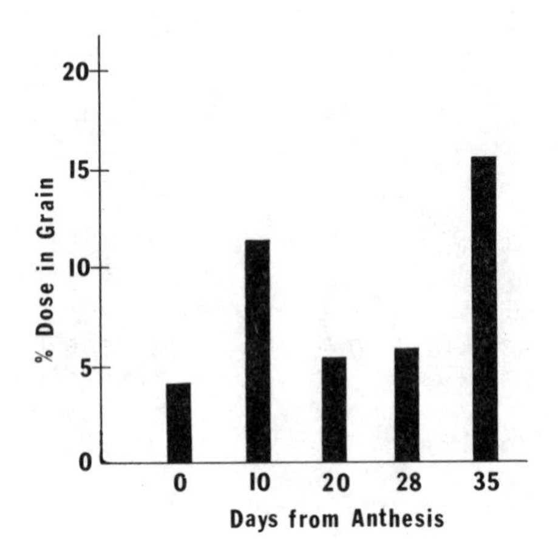

Figure 1. ^{75}Se accumulation efficiency in wheat vs day dosed.

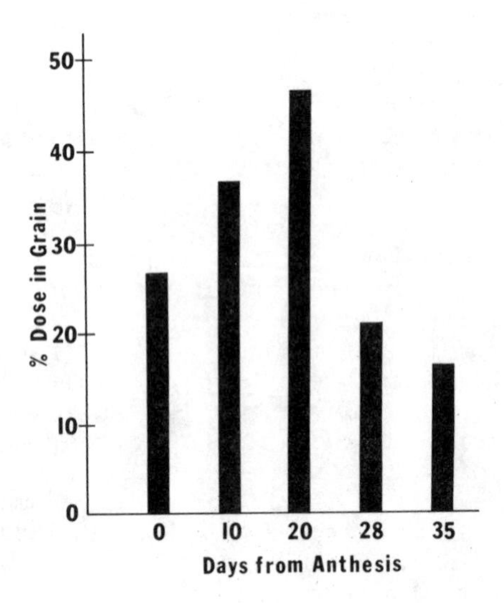

Figure 2. ^{65}Zn accumulation efficiency in wheat vs day dosed.

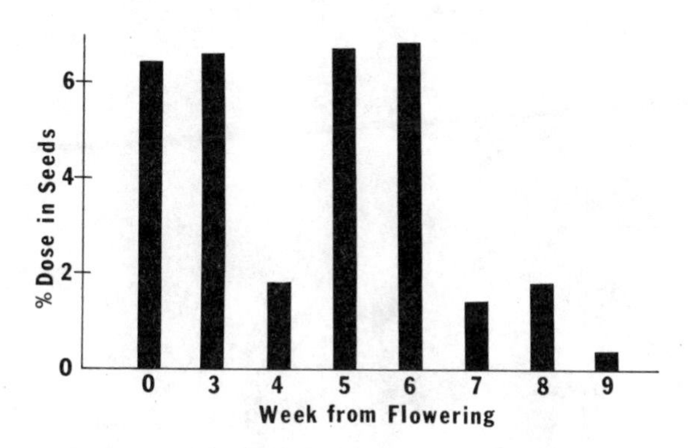

Figure 3. ^{75}Se accumulation efficiency in soybeans vs day dosed.

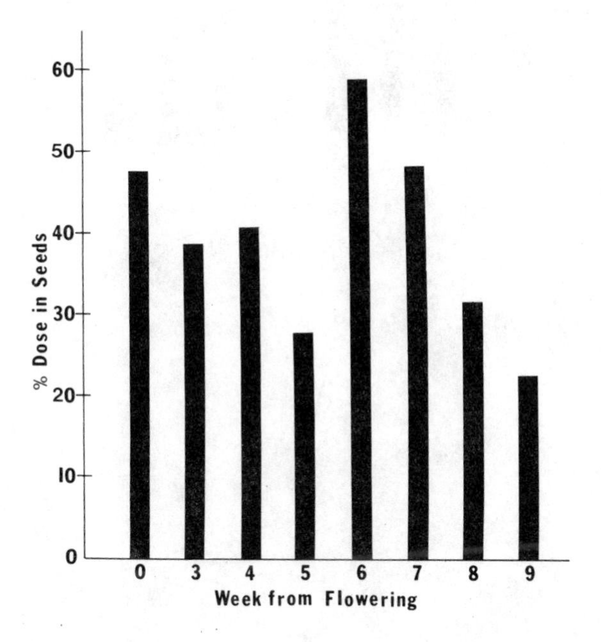

Figure 4. ^{65}Zn accumulation efficiency in soybeans vs day dosed.

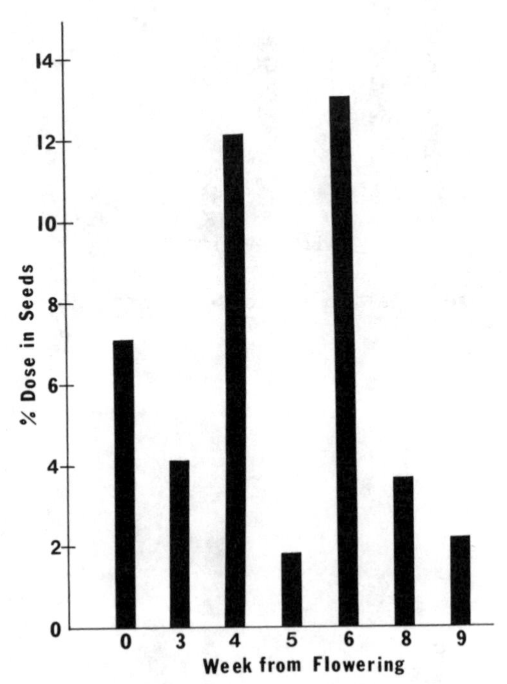

Figure 5. ^{59}Fe accumulation efficiency in soybeans vs day dosed.

Table IV. Concentration of Radionuclide in Mature Soybean Seeds as Related to Stage of Development When Dosed

Stage of Development When Dosed[1]	Concentration of Radionuclide ^{65}Zn cpm/g	^{75}Se cpm/g	^{59}Fe cpm/g
Beginning flowering		14,400	
End flowering	45,769	16,215	6,787
Small pod	49,759	13.023	4,631
Seed developing	34,357	11,735	5,425
Seed almost full sized	37,367	10,367	5,307

[1] Radionuclides were applied as a single dose 5 weeks from first flowering.

Table V. Concentration of Radionuclide in Mature Wheat Grains as Related to Stage of Development When Dosed

Stage of Development When Dosed	Concentration of Radionuclide ^{65}Zn cpm/g	^{75}Se cpm/g
4 wks before anthesis		31,792
3 wks before anthesis		26,980
2 wks before anthesis	331,009	36,487
1 wk before anthesis	298,695	32,047
2 wks after anthesis	478,440	17,518
3 wks after anthesis	454,831	22,949
3.5 wks after anthesis	484,431	29,646

[1] Radionuclides were applied as a single dose 3.5 weeks after beginning anthesis.

from one period of radionuclide application for either soybeans or wheat are presented because no obvious differences were observed between radionuclide accumulation when dosed prior to fertilization or during flowering vs when seeds were maturing. One exception was a dramatic reduction in radionuclide concentration of soybeans dosed after pods began to senesce (not shown).

Production of Soybeans Labeled with ^{70}Zn

The radiotracer studies indicated that our system of labeling plants with a single dose of zinc could result in the deposition of a substantial portion of the dose in the seeds, especially if administered during flowering. Therefore, we undertook a study to label soybeans with the stable isotope ^{70}Zn (10). The percent of the applied dose accumulated by the seeds was 14-23 for 10 pots analyzed compared to the 27.6% ^{65}Zn accumulated by soybean seeds treated similarly. The natural mass isotope ratio of ^{70}Zn:^{68}Zn in the soybeans was enriched by 85-fold (Table VI). This level of enrichment is sufficient for use in human bioavailability studies using 100 g portion of the labeled seeds per person.

Table VI. Soybeans Labeled With ^{70}Zn[1]

Mass Isotope Ratio	$^{70}Zn/^{68}Zn$
Natural	0.0343
Applied Isotope	8.05
Seeds	2.93$\pm$0.38

[1] See Ref. (10) for details.

Maximal enrichment of plant tissues with stable isotopes would occur if the mineral in the nutrient solution were totally replaced with the stable isotope throughout the entire growing period. However, this would not be the most efficient use of stable isotopes and could be prohibitively expensive for some stable isotpes such as ^{70}Zn in quantities needed for a human bioavailability study.

Acknowledgments

The work reported here was supported in part by a grant from USDA (Competitive Research Grants Office, CSRS, S&E, 59-2253-7-7-769-0). The technical assistance of Nancy Meyer, April Mason, and Stephen Lord is gratefully acknowledged.

Literature Cited

1. Cook, J.D.; Layrisse, M.; Martinez-Torres, C; Walker, R; Monsen, E; Finch, C.A. J. Clin. Invest. 1972, 51, 805-815.
2. Bjorn-Rasmussen, E.; Hallberg, L.; Walker, R.B. Am. J. Clin. Nutr. 1973, 26, 1311-1319.
3. Evans, G.W.; Johnson, P.E. Am. J. Clin. Nutr. 1977, 30, 873-878.
4. Meyer, N.R.; Stuart, M.A.; Weaver, C.M. J. Nutr. 1983, 113, 1255-1264.
5. Janghorbani, M.; Istfan, N.W.; Pagounes, J.O.; Steinke, F.H.; Young, V.R. Am. J. Clin. Nutr. 1982, 36, 537-545.
6. Hoagland, D.R.; Arnon, D.I. "The waterculture method for growing plants without soil", Cal. Agr. Exp. Sta. Circ. 347 1950.

7. Weaver, C.M.; Schmitt, H.A.; Stuart, M.A.; Mason, A.C.; Meyer, N.R.; Elliott, J.G. J. Nutr., In Press.
8. Baker, D.A. in "Metals and Micronutrients: Uptake and Utilization by Plants"; Robb, D.A.; Pierpoint, W.S., EDs.; Academic: New York, 1983; Chap. 1.
9. Tiffin, L.O. in "Micronutrients in Agriculture"; Mortvedt, J. J.; Giordano, P.M.; Lindsay, W.L., Eds.; Soil Science Society of America, Inc.: Madison, WI, Chap. 9.
10. Janghorbani, M.; Weaver, C.M.; Ting, B.T.G.; Young, V.R. J. Nutr. 1983, 113, 973-978.

RECEIVED January 20, 1984

6

Stable ^{26}Mg for Dietary Magnesium Availability Measurements

RUTH SCHWARTZ

Division of Nutritional Sciences, Cornell University, Ithaca, NY 14853

Magnesium has only one short lived radiotracer ^{28}Mg (T 1/2=21.3 h). Thus stable ^{26}Mg (11.01%) offers not only a radiation-free alternative to ^{28}Mg, it can be used as a second, non-decaying tracer in single and dual tracer studies on magnesium metabolism. So far, only a fraction of the potential of ^{26}Mg has been realized. Magnesium-26 can be analyzed by neutron activation analysis (NAA) as well as by various mass spectometric techniques. Mass spectrometers which are most accessible to investigators in the biological sciences utilize electron impact ionization (EIMS). This has been the preferred analytical technique for ^{26}Mg in nutritional studies to date since it is more sensitive and precise than NAA. The usefulness of ^{26}Mg is limited by its relatively high natural abundance. Magnesium-26 doses needed for its accurate detection in blood or tissues are too large for injection into the circulation. For this reason, research using ^{26}Mg has been limited to studies of Mg absorption and bioavailability. Magnesium-26 has been intrinsically incorporated into leafy vegetables. These were subsequently labelled extrinsically with ^{28}Mg and tested for isotopic exchangeability in rats and in human subjects. In addition, true Mg absorption was measured by simultaneous administration of ^{26}Mg (orally) and ^{28}Mg (intravenously). The relative merits and limitations of ^{26}Mg and ^{28}Mg are discussed with emphasis on their utilization in dual tracer protocols.

Magnesium is the fourth most abundant element in the body and, after potassium, the most abundant intracellular cation. It is a co-factor or participant in numerous biological and physiological processes, encompassing such diverse functions as hydrolysis and

0097-6156/84/0258-0077$06.00/0

synthesis of ATP and all enzymatic reactions that depend on ATP, maintenance of membrane integrity, control of muscle tone and contraction, and the functioning of the nervous system (1). Despite the well established importance of magnesium in living systems, information is scant on its various roles in the body and their relationships to health and disease. The Recommended Dietary Allowances formulated by the National Research Council have included recommendations for magnesium since 1968 (2), but the levels recommended are still under debate (3). There are few reliable indicators of marginal magnesium status, and little is known with certainty of the adequacy of magnesium in normal diets.

One of several factors that have impeded research on magnesium metabolism is lack of a satisfactory isotope. The short-lived radiotracer ^{28}Mg ($T_{1/2}$ = 21.3h) (4,5) has been used in intact animals and man, primarily, to estimate rapidly exchanging magnesium pools (6-11). Such studies, which seldom exceed 24-48 hours in duration, have yielded useful information in magnesium-depleted patients before and after magnesium therapy (10,11). However, even with high initial doses of ^{28}Mg that allow radioactivity measurements to be continued for up to six days, less than 20% of body magnesium can be accounted for by isotope dilution (9). The short half life of ^{28}Mg further limits estimations of magnesium absorption which may require fecal collections for longer periods than those optimal for ^{28}Mg detection (9,12,13) by methods other than whole body counting (13). Clearly, a second magnesium isotope, not limited by rapid decay, would be useful in research on magnesium metabolism.

The following is an examination of the potential of the stable isotope ^{26}Mg (11.01%) as an alternative and complement to ^{28}Mg. While the second minor isotope of magnesium, ^{25}Mg (10.00%), has a lower natural abundance, ^{26}Mg has the advantage of being detectable by analytical techniques that are currently accessible to investigators in biomedical fields (14-16). These techniques have been published elsewhere (14,15) and will not be described in detail here except where such details serve to underline or illustrate the central theme of this review: to compare and contrast tracer capabilities of ^{26}Mg and ^{28}Mg.

Methods for the Detection of ^{26}Mg

The most accurate and precise method for isotope ratio measurements of metals available at present is thermal ionization mass spectrometry (TIMS). The technique has been used successfully for isotope dilution measurements of magnesium (17), but no tracer studies with magnesium isotopes using TIMS have been published to date. Thermal ionization mass spectrometry is fairly time consuming, even with recently developed automated instruments (18). This factor must be weighed against the superior precision of TIMS, however, which greatly reduces the need for multiple replicates. By far the greatest deterrant to the use of TIMS in in vivo tracer studies with stable isotopes is the cost of the instruments which are seldom accessible to investigators in biomedical research units.

Two methods that have been used in tracer studies with ^{26}Mg are neutron activation analysis (NAA) and EIMS, mass spectrometry of a volatile chelate with electron impact ionization (15,19,20). Neither technique can match TIMS in accuracy or precision, but both may offer greater sample throughput. Until more precise techniques become more generally available, research utilizing stable metal isotopes will remain limited to the conditions imposed by the analytical facilities at hand. As will be shown below, research with stable magnesium isotopes to date, has had to be tailored to the limitations of available analytical techniques.

Neutron Activation Analysis. Magnesium-26 has a small cross section of 0.03 b. The product of irradiation with thermal neutrons is ^{27}Mg ($T_{1/2}$ = 9.5m). As shown in Table 1, several elements commonly present in biological materials give rise to radioactive nuclides with radiations at energy levels close to those characteristic of ^{27}Mg. Neutron activation was used in the first trials of ^{26}Mg as an in vivo tracer when measurements were made with a well-type NaI-Tl crystal detector (14,21). Under these conditions the presence of sodium, aluminum and manganese in the samples interfered in the accurate detection of ^{26}Mg, but could be reduced or eliminated by sample purification.

Table I. NAA Characteristics of ^{26}Mg and Common Contaminants

Metal	Natural Abundance %	Cross Section (b)	Product Nuclide[1]	$T_{1/2}$	Major Energies (MeV)
^{26}Mg	11.01	0.038	^{27}Mg	9.5m	0.84 (70), 1.013
^{27}Al	100	0.235	^{28}Al	2.2m	1.78 (100)
^{23}Na	100	0.400	^{24}Na	15h	1.37 (100), 2.75
^{40}Ar	99.6	0.610	^{41}Ar	1.8h	1.29 (100)
^{55}Mn	100	13.3	^{56}Mn	2.6h	0.847 (99), 1.81

[1] Irradiation with thermal neutrons

Samples were purified before activation using the solvent extraction procedure of Hahn *et al.* (22) with thenoyltrifluoroacetone as the ligand. An activation routine was developed to avoid subsequent contamination with aluminum, to allow the excape of ^{41}Ar, to monitor flux variations, and to facilitate spectrum stripping of residues of manganese, aluminum, and sodium (15). The advent of high efficiency germanium detectors has significantly reduced the influence of contaminants other than ^{56}Mn, which has a far longer half life than ^{27}Mg. Corrections for ^{56}Mn can be made by repeated counting after ^{27}Mg has decayed to non-detectable levels.

Neutron activation analysis has low sensitivity for the detection of ^{26}Mg. The best results were obtained, in our hands, by irradiating samples containing 200-300 μg natural magnesium (20-35 μg ^{26}Mg) for one minute at a neutron flux of $10^{12} cm^{-2} sec^{-1}$ followed by detection with a HP-Ge detector. Since it is usually necessary to run activations in triplicate and to carry out additional analyses for total magnesium by an independent method, such as atomic absorption, NAA is not suitable for specimens containing magnesium at low concentrations. It was found suitable for the analyses of fecal samples, marginal for urine and unsuitable for plasma (15,21).

Mass Spectrometry with electron impact ionization (EIMS). Mass spectrometers with electron impact ionization are widely used in biomedical research, primarily for identification and quantitation of organic molecules. Adaptation of EIMS to isotope ratio measurements of minerals has, therefore, considerable appeal for investigators interested in using stable mineral isotopes as biological tracers.

Since EIMS is suitable only for relatively volatile materials, most metals must be converted to volatile chelates before they can be analysed by this technique. The magnesium chelate found most amenable to EIMS is the diketonate Mg(2,2',6,6'-tetramethyl-3,5-heptanedione)$_2$ (Mg(THD)$_2$) (19). The chelate can be formed in aqueous solutions at pH >9 in the presence of excess THD (16). Extraction into ethyl ether provided a simple method for separating the chelate from excess THD which otherwise interferes in the MS analysis of Mg(THD)$_2$. The ether layer was transferred to a 10 ml glass tube and allowed to stand at room temperature. As the ether evaporated, crystals of Mg(THD)$_2$ were deposited on the middle and upper inside surfaces of the tube, leaving a THD layer in the well. The latter was aspirated and washed out with 2 successive 200-300 μl aliquots of methanol. The crystals were redissolved in methanol at a concentration of 1-4 mg ml^{-1} Mg(THD)$_2$. Recovery of Mg as the chelate is about 25-45%.

Reproducible mass spectra could be obtained by introducing 2-4 μg of the chelate by direct probe into a Finnigan 3300 quadrupole mass spectrometer using a heating rate of 200° per hour (16). Under these conditions, the most abundant ion peaks seen were those of $MgTHD^+$ with peaks at m/z 207, 208, 209 corresponding to ^{24}Mg, ^{25}Mg, and ^{26}Mg respectively. Carbon-13 contributions from ^{24}MgTHD raise the relative intensity at m/z 208 to about twice the level that can be accounted for by the natural abundance of ^{25}Mg, while the peak at m/z 209 is increased by less than 8% by ligand contributions of ^{13}C and ^{18}O. The relative intensities determined in standards of known isotope concentrations deviated somewhat from theoretical values and varied from day to day. Consequently, all isotope ratio measurements in unknown mixtures were based on standard determinations included in each set of analyses.

Table II. Comparison of NAA and EIMS Analyses for ^{26}Mg.

	NAA	EIMS
μg Mg per replicate	100	0.1-0.2
Detection limit, % excess of n.a.[1]	15-20	5
% precision at detection limit	26	18
Maximum precision, (% S.D.)	3-4	0.5

[1] In excess of natural abundance

Table II summarizes the differences in EIMS and NAA for the detection of ^{26}Mg in biological materials. Sample processing for EIMS analysis is simpler and more rapid than that in preparation for NAA. The absolute sensitivity is increased by about two orders of magnitude, and precision by a factor of 2-3. Consequently, lower levels of ^{26}Mg excess can be detected with confidence (16). This comparison suggests that EIMS can be applied to ^{26}Mg measurements in speciments likely to be low in total magnesium and ^{26}Mg excess, e.g. urine and plasma. Unfortunately, these expectations were not fully realized.

The comparisons summarized in Table II do not take into account the influence of instrumental memory, a factor that was observed during earlier trials (15,16) but became increasingly troublesome as the volume of analyses increased. The effect of memory can be minimized by grouping sample sets within incremental ranges of 10-20% in excess of natural abundance, and by adapting the mass spectrometer with standards enriched with ^{26}Mg to the levels anticipated in a given series of analyses. Four to six replicates were analysed routinely, but the number of replicates needed was greater with samples highly enriched with ^{26}Mg. Theoretically, the precision with which isotope excess is measured increases with increasing levels of enrichment. Because of the memory problem, precision expressed as % standard deviation was best at enrichment levels of 20-35% in excess of natural abundance. At these levels the relative standard deviation for recovery of added ^{26}Mg was 3-5% of the mean. At higher enrichment levels NAA may provide better precision than EIMS. But, as will be shown below, most enrichment levels encountered in ^{26}Mg tracer studies are less than 40% in excess of natural abundance. It must be stressed that both methods have deficiencies. Until better techniques become available the method chosen by investigators will continue to depend on the facilities available to them.

Potential Uses of ^{26}Mg as an alternative or Complementary Tracer to ^{28}Mg

The obvious advantages of a stable tracer over a short-lived radioactive isotope are lack of radiation hazards, flexibility in timing of experiments independantly of isotope production schedules, and the possibility of storing for indefinite periods materials enriched with the tracer. These features add possibilities to research to magnesium metabolism that could not be considered when ^{28}Mg was the only available tracer isotope, if ^{26}Mg can, in fact, replace or complement ^{28}Mg as a tracer. The extent to which this is possible will be examined by considering the major uses of mineral tracers in metabolic research for: kinetic studies, determination of absorption patterns, measurements of true absorption, endogenous fecal excretion, and bioavailability from dietary sources.

Kinetic Studies. Magnesium-26 has a relatively high natural abundance and thus does not satisfy one of the major requirements for a true tracer: that it be detectable in systems to which it has been added in amounts that do not measurably alter the steady state. Table III shows that ^{28}Mg and ^{26}Mg can, theoretically, be detected with fairly comparable precision in the plasma of a man for 12-24 hours after an injection of 20 µCi ^{28}Mg or 20 mg ^{26}Mg. However, while 20 µCi of virtually carrier-free ^{28}Mg have no discernible effect of plasma magnesium pools, 20 mg ^{26}Mg are likely to increase total plasma magnesium by 25-35% and unbound magnesium by about 80%. Kinetic analysis based on the resulting plasma isotope curve would lead to erroneously high estimates of exchangeable magnesium pools. Clearly, ^{26}Mg is not suitable for kinetic studies or any other purpose such as estimation of endogenous fecal magnesium for which it must be administered by injection.

Table III. Plasma Isotope Concentration after an I.V. Injection of 20 µCi ^{28}Mg or 20 mg ^{26}Mg

Time (h)	^{28}Mg			^{26}Mg	
	% Dose L^{-1}	cpm[1]	% S.D.	% e[2]	% S.D.
0.5	4.6	4020	<1	42	1.8
4	1.7	1300	1.3	25	2.8
8	1.1	850	2.5	10	4.5
12	0.9	500	3.8	8	5.5
24	0.7	300	4.5	6	7.5
48	0.5	90	11.0	4.5	15.0

[1] In 4 ml plasma at the time sample was taken, not corrected for decay.
[2] % in excess of natural abundance.

Estimation of Magnesium Absorption Patterns. The determination of calcium absorption patterns based on plasma or urinary isotope levels measured within the first 12 hours after isotope ingestion has useful diagnositc applications (23,24). A similar test has not been described for magnesium although, increasingly, defects in magnesium absorption are being recognized in clinical practice (25). Since such tests are usually carried out in the fasting state, it is possible to administer fairly large doses of ^{26}Mg without appreciable dilution with natural magnesium. Table IV shows urinary levels of ^{26}Mg excess determined in a patient who had been given 50 mg ^{26}Mg as part of a breakfast meal which contained about 50 mg natural magnesium. Tolerable levels of precision were achieved with measurements of ^{26}Mg excess in urine composites collected over 0-4, 4-8, and 8-12 hours after the labelled breakfast. Doses of up to 100 mg ^{26}Mg are feasible, particularly if the remainder of the meal is low in magnesium, for a simple test of magnesium absorption capacity not requiring prolonged periods of collection or observation.

Table IV. Isotope Levels in Urine Following 20 µCi ^{28}Mg I.V. or 50 mg ^{26}Mg P.O.

Time	% Dose L^{-1}	^{28}Mg cpm[1]	% S.D.	% Dose L^{-1}	^{26}Mg % e[2]	% S.D.
0-4	5.0	26,000	<1	1.1	24	3
4-8	2.0	9,000	<1	0.9	14	5
8-12	1.3	4,100	<1	0.7	7	10
12-24	1.3	2,200	>1	0.7	6	13
24-48	2.7	670	3.5	1.4	4	26

[1] In 15 ml urine at the time of sample collection, not corrected for decay.
[2] % in excess of natural abundance.

Measurements of True Magnesium Absorption. Appearance of a single ingested isotope in the plasma or urine can not provide a quantitative estimate of fractional absorption of a mineral without concomitant measurement of isotope exchange between the mineral in the plasma and other body compartments (24). Table IV shows the levels of urinary ^{28}Mg from an intravenous injection that accompanied an oral dose of 50 mg ^{26}Mg. Radioactivity was determined in 15 ml urine with a relative S.D. of < 1% in composites collected at intervals up to 24 hours, and with < 4% in the 24-48 h collection. The relative standard deviations for ^{26}Mg measurements were consistently higher, reaching unacceptable levels in samples taken after 12 hours following the test meal.

"True" fractional absorption can be determined from urinary isotope measurements by estimating the ratio $SA_o : SA_i$ where Sa is the specific activity of the urine as % Dose of tracer per unit of the mineral of interest and the suffixes p,i refer to oral and intravenous respectively (25). Because the initial declining portions of the oral and intravenous tracer curves differ. (Figure 1), the ratio is best determined on the terminal parallel portions of the curves, or as far removed in time from the point of isotope administration as possible. As the summary in Table V shows, ratios determined in urine samples taken after 24 hours may be less acceptable than those based on ratios in earlier samples which, however, are likely to overestimate "true absorption". The precision of the test could be enhanced by increasing the dose of ^{26}Mg, but since fractional magnesium absorption varies inversely with the quantity consumed, the stable isotope dose should not exceed the amounts found in normal meals, at most 100-150 mg Mg.

Table V. Ratios ^{26}Mg: ^{28}Mg In Urine At Increasing Intervals After 20 μCi ^{28}Mg I.V., 50 mg ^{26}Mg P.O.

Time (h)	Mean[1]	Ratio Low	High
4-8	0.45	0.43	0.47
8-12	0.54	0.48	0.54
12-24	0.54	0.46	0.61
24-48	0.52	0.40	0.64
	0.51	0.44	0.57

[1] Mean of three replicate analyses

"True" absorption can also be determined by correcting net absorption for endogenous fecal excretion (26). Net absorption is measured by subtracting fecal isotope excretion from isotope intake. It is by far the most reliable method for fractional absorption measurements with ^{26}Mg, given the limitations of ^{26}Mg detection by EIMS.

Table VI shows ^{26}Mg excess levels in daily fecal excretions and in composites of 5 and 9-day pools. The subject illustrated in Table VI showed an unusually rapid return of the isotope dose. In this instance a fecal collection period of 5-6 days should have been adequate for recovery of at least 95% of the ingested tracer. Since some subjects may require longer fecal collections, polyethyleneglycol (PEG) 4000 was included with test meals (Table VII). The results showed a constant relationship of ^{26}Mg to PEG in all fecal pools that contained at least 40% of ingested PEG. It should thus be possible to make accurate measurements of fractional magnesium isotope absorption by limiting isotope analyses to the first four or five days of fecal excretion when isotope concentrations are at levels detectable with acceptable precision.

^{26}Mg in Measurements of Mineral Bioavailability from Dietary Sources. One of the most convenient methods for measuring dietary mineral bioavailability is to follow the absorption of a tracer

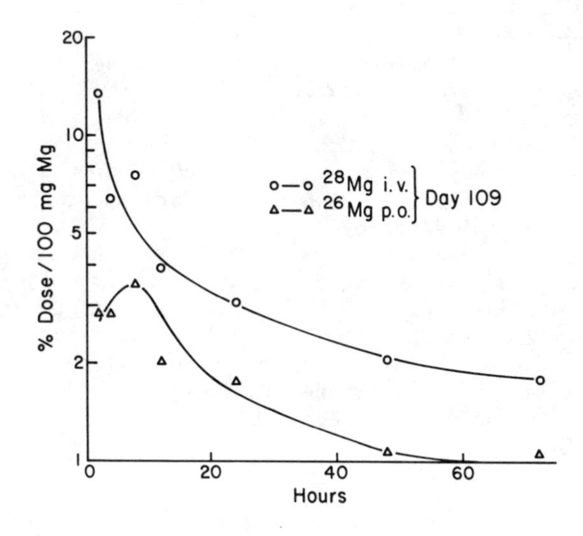

Figure 1. Urinary Excretion of ^{28}Mg (i.v.) and ^{26}Mg (p.o.) administered to a subject who had consumed a constant adequate diet for 109 days. Isotope doses were 20μCi ^{28}Mg, 50 mg ^{26}Mg.

Table VI. Fecal Isotope concentrations after 30 μCi ^{28}Mg and 50 mg ^{26}Mg (orally)

Time Days	^{28}Mg % Dose	cpm[1] (15 ml)	% S.D.	^{26}Mg % e.[2]	% S.D.
1	45.5	348,000	<0.1	143	0.6
2	8.4	29,000	<1.0	25	2.8
3	0.8	1,300	3.0	3	27
4	0.3	250	>5.0	1.4	50
5	0.2	80	> 20	0.9	125
1-5	55.0	4,300	1.5	42	2.3
1-9	"	270	5.0	16	4.5

[1] Ave. volume of fecal homogenate = 300 ml day^{-1}
[2] % in excess of natural abundance

isotope which has been added extrinsically to the food or meal of interest. This method is only valid, however, if the tracer can be assumed to mix completely with the dietary mineral before it arrives at its normal sites of absorption. The validity of this assumption can be ascertained by intrinsically (biologically) labelling foods and subsequently comparing the fractional absorption of the intrinsic tag with that of a second isotope that is added extrinsically to the same food. The test is best carried out with two isotopes used at the same time (27) but can be done with a single isotope administered in sequence intrinsically and extrinsically to the same subject.

Table VII. ^{28}Mg: ^{26}Mg Exchangeability and Net Absorption[1] from Bran[2] and Vegetable[3] Muffins

Test Meal	Net Absorption ^{26}Mg	$^{26}Mg:^{28}Mg$
	Year 1	
Bran	41.6±7.1[a]	1.1 ±0.18
Collards	54.2±6.5[c]	0.95±0.94
Spinach	47.5±5.6[b]	0.95±0.05
	Year 2	
Bran	41.9±8.2[a]	1.05±0.13
Lettuce	51.2±7.9[b]	0.98±0.08
Turnips	53.7±7.4[c]	0.97±0.11

[1] $\frac{\text{\% Dose PEG} - \text{\% Dose Isotope}}{\text{\% Dose PEG}}$ (all measurements in feces).
[2] Standard test meal(see Table VIII).
[3] Intrinsically labelled with ^{26}Mg.
[a,b,c] Means without a common letter in the superscript are significantly different ($p < 0.05$). Comparisons were made vertically and horizontally.

Magnesium-26 offers the opportunity for the applications of this procedure in magnesium bioavailability studies that were not feasible with the short lived ^{28}Mg. Four leafy vegetables were intrinsically labelled by using ^{26}Mg as the sole source of magnesium in nutrient culture solution (28). The vegetables were then extrinsically tagged with ^{28}Mg and tested in rats (28) and in human volunteer subjects (29). Table VII shows the data obtained in the tests in human subjects. Exchangeability of the extrinsic and intrinsic tracers was close to 100% in both species. Fractional magnesium absorption was similar in all test vegetables and of the same order in rats (28) as in the human subjects (29).

Table VIII. Magnesium and ^{26}Mg Content of Test Meals (3 Muffins) (29)

	% ^{26}Mg in:		mg per 3 muffins:	
Test Meal	Vegetable	Muffin	Total Mg	^{26}Mg added as tracer
Standard Bran	--	11.0	148	48[1]
Collards	90.9	56.0	149	50[2]
Lettuce	92.0	48.6	156	42
Spinach	89.0	52.6	153	46
Turnips	87.4	47.2	154	39

[1] Added extrinsically in solution to replace the daily supplement of 50 mg Mg.
[2] In the muffins. The supplement of 50 mg Mg (as MgO) was taken with test meal.

The test meals consumed by the human volunteers were bran muffins (Table VIII) in which about half the bran was replaced by a ^{26}Mg-labeled freeze-dried vegetable. The standard test meal to which both isotopes were added in solution as $MgCl_2$, therefore, contained about twice the amount of wheat bran included in the vegetable muffins. As Table VII shows, fractional magnesium absorption was consistently and significantly lower from the standard test meal than from the meals containing one of the green vegetables, suggesting that bran may suppress magnesium availability.

Conclusions regarding the potential of ^{26}Mg as a substitute and complement for ^{28}Mg

At present, uses of ^{26}Mg are restricted to objectives that can be attained by oral isotope administration only. One of these is the measurement of "true" magnesium absorption in which ^{26}Mg is given orally, and ^{28}Mg by injection. "True" absorption is then computed either by estimation of the ratio ^{26}Mg:^{28}Mg in urine or, more precisely, by measuring net absorption (subtraction of fecal excretion from isotope intake) corrected for endogenous fecal excretion. The latter is determined with data obtained with intravenously injected ^{28}Mg (26). A second, and probably the most useful application of ^{26}Mg to date, is in measurements of dietary magnesium bioavailability in which ^{26}Mg is used as an extrinsic label. Such measurements can be carried out with ^{26}Mg alone, using each subject as his own control in a sequence of tests comparing net magnesium absorption from different foods, or to assess the effect of certain dietary components, e.g. fibre, phytates or other minerals on Mg absorption.

Acknowledgments

Supported in part by NIH grant #18569 and USDA Cooperative Agreement #12-44-1001-814.

Literature Cited

1. Wacker, W. E. C. "Magnesium and Man"; Harvard University Press: Cambridge, Massachusetts, 1980.
2. Recommended Dietary Allowances (7th Edition); N. R. S.; Nat'l. Acad. Sci.: Washington, D.C., 1968.
3. Seelig, M. A. "Magnesium Deficiency in the Pathogenesis of Disease"; Plenum Publishing Corp: New York, 1980; p. 9.
4. Sheline, R. K.; Johnson, N. R. Phys. Rev., 1953, 89, 520-521.
5. Jones, J. W.; Kohman, T. P. Phys Rev., 1953, 90, 495-496.
6. Aikawa, J. K.; Gordon, G. S.; Rhoades, E. L. J. Appl. Physiol., 1960, 15, 503-507.
7. Silver, L.; Robertson, J. S.; Dahl, L. K. J. Clin Invest., 1960, 39, 420-425.
8. Graham, L. A.; Casear, J. J.; Burgen, A. S. V. Metab., 1960, 9, 646-.
9. Avioli, L. V.; Berman, M. J. Appl. Physiol., 1966, 21, 1688-1964.
10. Wallach, S.; Dimich, A. Ann. N. Y. Acad. Sci., 1969, 162, 963-973.
11. Jones, J. E.; Shane, S. R.; Jacobs, W. H.; Flink, E. B. Ann. N. Y. Acad. Sci., 1969, 162, 934-946.
12. Roth, P.; Werner, E. Int. J. Appl. Rad. Isotopes, 1979, 30, 523-526.
13. Watson, W. S.; Hilditch, T. E.; Horton, P. W.; Davies, D. L.; Lindsey R., Metab. 1979, 28, 90-123.
14. Currie, V. E.; Lengemann, F. W.; Wentworth, R. A.; Schwartz, R. Int. J. Nucl. Med. Biol., 1975, 2, 159-164.
15. Schwartz, R.; Giesecke, C. C. Clin. Chim. Acta, 1979, 97, 1-8.
16. Schwartz, R.; Spencer, H.; Wentworth, R. A. Clin. Chim. Acta, 1978, 87, 265-273.
17. Gardner, E. L.; Machlan, L. A.; Gramlich, J. W.; Morre, L. J.; Murphy, T. J.; Barnes, I. L. NBS Publication, 1974, 422, 951-960.

18. Janghorbani, M.; Young, V. R.; Gramlich, J. W.; Machlan, L. A. Clin. Chim. Acta, 1981, 114, 163-171.
19. Schwartberg, J. E.; Sievers, R. E.; Moshier, R. W. Anal. Chem., 1970, 42, 1828-1830.
20. Hachey, D. L.; Blais, J. C.; Klein, P. D. Anal. Chem., 1980, 52, 1131-1135.
21. Schwartz, R. Fed. Proc., 1982, 41, 2709-2713.
22. Han, K. J.; Tuma, D. J.; Quaife, M. A. Anal. Chem., 1967, 39, 1169-1170.
23. Raymakers, J. A.; Roelofs, J. M. M.; Duursma, S. A.; Visser, W. J. Neth. J. Med., 1975, 18, 191-204.
24. Roelofs, J. M. M.; Raymakers, J. A. Clin. Chim. Acta, 1976, 67, 53-62.
25. Flink, E. B. in Prasad, A., Ed. "Trace Elements in Human Health and Disease", Acad. Press, New York, 1976, p. 4.
26. Comar, C. L. "Radioisotopes in Biology and Agriculture", McGraw-Hill: New York, 1955; p. 14.
27. Monsen, W. R.; Hallberg, L; Layrisse, M.; Hegsted, D. M.; Cook, J. D.; Mertz, W.; Finch, C. A. Amer. J. Clin. Nutr., 1978, 31, 134-141.
28. Schwartz, R.; Grunes, D. L.; Wentworth, R. A.; Wien, E. M. J. Nutr., 1980, 110, 1365-1371.
29. Schwartz, R.; Spencer, H.; Walsh, J. E., Amer. J. Clin. Nutr., 1983, in press.

RECEIVED January 31, 1984

7

GC/MS Measurement of Stable Isotopes of Selenium

For Use in Metabolic Tracer Studies

CLAUDE VEILLON

U.S. Department of Agriculture, Beltsville Human Nutrition Research Center, Building 307, Room 215, Beltsville, MD 20705

Numerous trace elements are known to be nutritionally essential in man. In order to assess the essentiality, dietary availability, and metabolic fate of these, means of labeling for subsequent identification are needed. In animal studies, radioisotopes are often used for this purpose, but their use in human studies is generally contraindicated due to the radiation hazards. An alternate method is to use stable isotopes of the elements, which overcomes this limitation. A method will be described for conveniently measuring the stable isotopes of selenium, permitting their use as metabolic tags in tracer studies. Using one stable isotope as the tracer and another as internal standard, one can quantitatively identify in a sample the tracer, natural (unenriched) selenium present with it, and total selenium. Some of the kinds of information obtainable from metabolic tracer studies will be discussed.

Selenium has been recognized as an essential trace element in the diets of man and animals for many years (1). Another strong indication of its essentiality is the fact that it is an essential component of the enzyme glutathione peroxidase (2). Recently, scientists from the People's Republic of China demonstrated that Keshan disease (a cardiomyopathy in children) was correlated with low dietary selenium intakes (3) and could largely be prevented with supplementation. Similarly, poor selenium nutrition in patients receiving total parenteral nutrition has been linked to muscular discomfort (4) and cardiomyopathy (5).

These studies indicate that a much better understanding of the role of selenium in human nutrition is needed. The types of questions to be answered more fully are the biological availability of selenium from foods, the fraction of dietary intake that is absorbed, the metabolic fate of absorbed selenium, bodily requirements, stores and nutritional status of individuals, and factors affecting these.

Stable isotopes of selenium (as well as those of other elements) can provide a means of addressing these questions by their employment in metabolic tracer studies. The same information can be obtained as when employing radiotracers in animal studies, but without the associated radiation hazards in human studies. For example, stable isotopes of selenium can be biologically incorporated into test foods and these used to monitor selenium bioavailability (6,7).

Described herein is a convenient, accurate, sensitive and rapid method for measuring stable isotopes of selenium in biological materials. A rapid sample preparation technique is used which does not require perchloric acid, with its associated dangers. Based on isotope dilution techniques and isotope ratio measurements, the method employs one enriched stable isotope as the internal standard and another as the metabolic tag. This permits the quantitative measurement of the enriched tracer, unenriched (natural) selenium present with it, and total selenium in the samples. Some examples of the type of information that can be obtained with these techniques will be described.

Isotope Dilution

The technique of stable isotope dilution permits one to determine very accurately the trace element content of a particular sample as well as using stable isotopes as metabolic tracers. The concept of isotope dilution for trace element determinations is both simple and elegant. A known quantity of an enriched stable isotope is added to the sample to be analyzed. By measuring the amount of that isotope (added) relative to another isotope (not added) of the element, one can easily calculate the amount of analyte present originally in the sample. Thus, the normal relative amounts of the two isotopes (natural abundance) have been altered, or "diluted".

Advantages. Let us assume for the moment that the requirements for most trace element analysis methods are also met here. These are: that a known amount of an enriched isotope of the element in question (spike) has been added to the sample; that subsequent chemical processing renders the spike and endogenous analyte in the same chemical form; and that contamination is under control. The second assumption, equilibration of analyte and spike, can be a critical one, in that lack of equilibration could contribute systematic errors.

One major advantage of stable isotope dilution methodology is that quantitative recovery of the analyte is not required. Both the analyte and the added spike are identical chemically, so incomplete recoveries will affect both analyte and spike in the same way. In this regard the enriched spike serves not only as an internal standard for the determination, but also constitutes an ideal internal standard.

Another major advantage of the method is that it has the potential of being an absolute method, that is, unlike most other methods, it is not necessary to calibrate instrument response against known standards. It is simply a ratio measurement, of two chemically identical species.

Limitations. About 17 elements are mononuclidic, i.e., only one isotope of the element exists in nature, and several of these are of interest biologically, such as F, Na, Al, P, Mn, Co, and As.

In studies where radioisotopes can be used, using stable isotopes is usually less convenient and more time consuming. For trace element determinations, other methods like atomic absorption spectrometry are usually faster, more convenient and require simpler instrumenation. Other than for tracer studies in humans, stable isotope dilution methods perhaps serve best in establishing the accuracy of other methods.

Basis of the Method

In this method, samples are spiked with a known amount of an enriched isotope of selenium (^{82}Se in this case) and wet digested to destroy the organic matter and render all of the selenium into the +4 oxidation state. Next, the selenium is reacted with 4-nitro-o-phenylenediamine (NPD) to form the nitropiazselenol (Se-NPD) which is extracted into chloroform. Aliquots of the extract are then introduced into the mass spectrometer (MS) via a gas chromatograph (GC) and individual ions measured to determine isotope ratios of the various selenium isotopes. From the observed isotope ratios and the known amount of ^{82}Se spike added, the amounts of enriched tracer (e.g., ^{76}Se, ^{74}Se, etc.), unenriched selenium and total selenium in the samples can be calculated. Let us now look at each of these steps in detail.

Sample Digestion. Samples (1-5 g or mL) are weighed or pipetted into 100 mL micro Kjeldahl flasks containing 2 glass boiling beads. Depending on sample size, 3-5 mL of concentrated HNO_3, 1 mL of H_3PO_4, and a known amount of enriched ^{82}Se are added and samples allowed to stand for 1 hr at room temperature. Then samples are heated to boiling and additions of 30-50% H_2O_2 begun slowly. This is continued until the volume is reduced to about 1 mL. Samples are then cooled briefly, 1 mL formic acid added, and reheated. This reduces any residual HNO_3 to NO_2. Next, 2 mL concentrated HCl are added and the samples boiled gently for 10 min to convert any Se (VI) to Se (IV). Ten mL of water are added and the samples allowed to cool. They are then extracted with 3-5 mL of chloroform to remove any lipids present.

Chelation. The nitropiazselenol is formed by adding 0.5 mL of 1% NPD in water and the samples allowed to stand for 1 hr at room temperature with occasional shaking. The resulting Se-NPD is then extracted into 2 mL chloroform with mechanical shaking for 15 minutes. The chloroform layer is removed, placed in small glass test tubes and evaporated in a vacuum oven (100 Torr) at 50°. Then 100 uL of chloroform are added, the tubes closed with polyethylene plugs, and aliquots of this solution (1-10 uL) injected into the GC/MS for analysis.

Total Selenium. As mentioned earlier, stable isotope dilution is a powerful tool in trace element analysis. Let us first look at how it can be used to determine the total selenium content of a sample. In the following section we will develop the method further for stable isotopes in metabolic tracer studies.

As it occurs in nature, selenium consits of six stable isotopes with the relative abundances shown in Table I.

Table I. Relative Abundance of Se Isotopes in Natural Se

Isotope	Abundance, atom %
74	0.87
76	9.02
77	7.58
78	23.52
80	49.82
82	9.19

When Se-NPD is introduced into the mass spectrometer, the most intense group of ions is that of the parent ion, Se-NPD$^+$. This is illustrated in Figure 1 for natural (unenriched) Se-NPD. Six peaks are observed, corresponding to Se-NPD$^+$ containing each of the six Se stable isotopes. By adding a known amount of enriched ^{82}Se (spike) to the samples, and monitoring the observed isotope ratio of, say, ^{80}Se/^{82}Se, we can readily calculate the amount of natural selenium that had to be present in the sample.

The enriched ^{82}Se spike which we use has relative abundances as shown in Table II.

Table II. Relative Abundance of Se Isotopes in Enriched ^{82}Se

Isotope	Abundance, atom %
74	0.13
76	0.19
77	0.30
78	0.60
80	1.96
82	96.81

Thus, we can set up an expression for the $^{80}Se/^{82}Se$ ratio as follows:

$$^{80}Se/^{82}Se = R = \frac{Se_n(0.498) + Se_{82}(0.0196)}{Se_n(0.0919) + Se_{82}(0.968)} \quad (1)$$

Here the numerator is the sum of all sources of ^{80}Se, i.e., 49.8% in natural selenium (Se_n) and 1.96% in the ^{82}Se (Se_{82}) spike (Tables I and II). Similarly, the denominator is all sources of ^{82}Se; 9.19% in Se_n and 96.8% in Se_{82}.

We wish to express the amounts of Se_n and Se_{82} as weights (e.g., ng). Yet the mass spectrometer measures the <u>number</u> of ions at a given mass. Since ^{82}Se and ^{80}Se differ in mass, a correction must be made to Equation 1. Looking at this another way, let us assume that we had added enough ^{82}Se to a sample so that the ^{80}Se and ^{82}Se peaks in Figure 1 were the same size. This would correspond to the same <u>number</u> of ^{82}Se and ^{80}Se ions but different <u>amounts</u> (weights) of each isotope. So, Equation 1 has to be modified:

$$R = \frac{Se_n(0.498) + 0.976\ Se_{82}(0.0196)}{Se_n(0.0919) + 0.976\ Se_{82}(0.968)} \quad (2)$$

The 0.976 factor now corrects this equation so that Se_n and Se_{82} are in terms of weights of each (80/82 = 0.976). In other words, for an equal number of ^{80}Se and ^{82}Se ions there is a greater weight of ^{82}Se present, so this term must be reduced by the 0.976 factor.

There is one more factor to consider here. We are not actually measuring $^{80}Se^+$ and $^{82}Se^+$, but $^{80}Se\text{-}NPD^+$ and $^{82}Se\text{-}NPD^+$. The NPD portion of the ion contains six carbon atoms, three nitrogen atoms and two oxygen atoms. Carbon is mostly ^{12}C, but 1.1% of carbon is ^{13}C. Likewise, nitrogen is 0.37% ^{15}N and oxygen is 0.20% ^{18}O. The occurrence of these in the ions will alter the $^{80}Se\text{-}NPD/^{82}Se\text{-}NPD$ ratio slightly from that of $^{80}Se/^{82}Se$.

The natural isotope ratio of $^{80}Se/^{82}Se$ is 5.42. Correcting for the contributions of the ligand isotopes, the ratio should be 5.36. This ratio is actually observed, indicating no instrumental bias in the measurements. Frew et al. (8) have discussed these corrections in detail. So, Equation 2 must be corrected for this small effect of the ratio differences of a factor of 1.01 (i.e., 5.42/5.36), namely:

$$1.01R = \frac{Se_n(0.498) + 0.976\ Se_{82}(0.0196)}{Se_n(0.0919) + 0.976\ Se_{82}(0.968)} \quad (3)$$

The correction for the ligand isotopes is small because of the 2 amu separation of the peaks. Adjacent pairs would have larger corrections.

Solving Equation 3 for Se_n, we get:

$$Se_n = \frac{Se_{82}(0.0191 - 0.954R)}{0.0928R - 0.498} \quad (4)$$

Here R is the observed ^{80}Se-NPD/^{82}Se-NPD ratio, Se_{82} is the known weight of the added ^{82}Se spike, and Se_n is the weight of natural selenium in the original sample in the same units as Se_{82}.

This relationship can also be used to establish the accuracy of the ^{82}Se spike concentration, which is most conveniently added as a solution. Employing accurately prepared samples of natural selenium (Se_n known) spiked with known amounts of the ^{82}Se solution, R is then measured and Equation 4 solved for Se_{82}. This is basically an inverse isotope dilution procedure. When performed on the same instrument to be used for subsequent measurements, it has the added advantage of cancelling out any mass discrimination by the instrument.

Tracer Studies. In addition to using ^{82}Se as an internal standard as described above, a second enriched isotope of selenium can be used as a metabolic tag. Let us take as an example the use of enriched ^{76}Se as a tracer. We have used a batch of ^{76}Se for this purpose, with the relative abundances shown in Table III.

Table III. Relative Abundances of Se isotopes in Enriched ^{76}Se

Isotope	Abundance, atom %
74	0.14
76	96.88
77	0.85
78	0.99
80	0.95
82	0.18

In this case, we will need to measure 2 isotope ratios, namely ^{80}Se/^{82}Se (spiking samples with ^{82}Se, as before) and ^{76}Se/^{82}Se (using ^{76}Se as the metabolic tag). Proceeding as before, we can set up expressions for these 2 ratios using the coefficients from Tables I-III:

$$R_1 = \frac{^{80}Se}{^{82}Se} = \frac{Se_n(0.498) + Se_{82}(0.0196) + Se_{76}(0.0095)}{Se_n(0.0919) + Se_{82}(0.968) + Se_{76}(0.0018)} \quad (5)$$

and:

$$R_2 = \frac{^{76}Se}{^{82}Se} = \frac{Se_n(0.0902) + Se_{82}(0.0019) + Se_{76}(0.969)}{Se_n(0.0919) + Se_{82}(0.968) + Se_{76}(0.0018)} \quad (6)$$

As before, we have to incorporate corrections for the ligand isotopes and mass differences, but we now have 2 equations with two unknowns which can be solved simultaneously. The algebraic manipulations are quite messy and will not be presented here. Many of the terms in the solution have small insignificant coefficients and can be eliminated. The solution simplifies to:

$$Se_n = \frac{Se_{82}(0.915R_1 - 0.009R_2 - 0.0184)}{0.482 - 0.090R_1} \quad (7)$$

and:

$$Se_{76} = \frac{R_2(0.085\ Se_n + 0.897\ Se_{82}) - 0.084\ Se_n - 0.0018\ Se_{82}}{0.0018R_2 - 0.9688} \quad (8)$$

From 2 ratio measurements and the known amount of ^{82}Se spike added, one can quantitatively identify: a) the ^{76}Se tracer present in the sample, and b) the amount of natural selenium present with it. As a bonus, one also gets the total selenium content of the sample, which is simply the sum of these.

Note that the equations derived here are valid only for the enriched materials described in Tables II and III. Different coefficients would have to be used for other enrichments and isotopes, but in principle any combination of isotopes could be used for internal standrad and metabolic tag.

Applications

Let us now consider some of the types of information one can obtain using stable isotopes in metabolic studies. In essence, one can obtain the same information as with radioisotopes, but with less convenient measurement. Stable isotopes also have the advantage of an "infinite" half-life, permitting long term studies. For an element like selenium with six stable isotopes, more than one can be used in the same experiment at the same time.

Several groups have been active in recent years in employing stable isotopes of various elements in tracer studies involving trace element metabolism. Janghorbani et al. (9-11) have recently reported on metabolic studies with Zn and Se stable isotopes. Carni et al. (12) investigated iron utilization using ^{58}Fe. In an elegant study, Harvey investigated Cu uptake in fish (13). Copper is an example of an element for which no suitably long-lived radioisotopes exist. Turnlund et al. (14) measured iron and zinc absorption in elderly men using ^{70}Zn and ^{58}Fe. Yergey and co-workers (15) have studied calcium metabolism with stable isotopes of that element, with relatively inexpensive instrumentation. Schwartz and Giesecke (16) investigated magnesium metabolism using both stable ^{26}Mg and the short-lived radioisotope ^{28}Mg. A number of elements of nutritional interest, including Mg, Ca, Cr, Fe, Ni, Cu, Cd and Zn, were investigated by Hachey and co-workers (17). Johnson (18) investigated mineral metabolism in human subjects employing ^{54}Fe, ^{57}Fe, ^{67}Zn, ^{70}Zn and ^{65}Cu. The method described herein has been described in part earlier (19). The methodology was subsequently employed in monitoring the intrinsic labeling of chicken products with ^{74}Se (7). These endogenously labeled products were subsequently used in a controlled metabolic study to measure selenium uptake in pregnant women (20).

Pool Sizes. Metabolic tracers can in principle be used to estimate the size of various body pools for trace elements. It is somewhat analogous to making specific activity measurements in radiotracer experiments.

Let us take as an example the following. We have a container of a fixed size or volume (P). Into this container water is flowing at a constant flow rate (U) and flowing out at the same rate through a radioactivity detector. Let us assume that the contents of the container are constantly and rapidly mixed. If at time t = 0 we introduce a known amount of soluble radioactive material into the inflowing stream, the concentration of this material in the outflowing stream would be at a maximum (rapid mixing assumed). The concentration would then decline as additional water flowed into the container, but in a predictable manner. It would decline exponentially and asymptotically approach zero as more water continued to flow in. If we were to plot counts-per-minute (CPM) in the outflow versus time we would see something like that shown in Figure 2. This process is known as exponential dilution and our hypothetical set-up would constitute an exponential dilution flask.

This is a general procedure and can be used to calibrate detector response in both liquid and gaseous systems. The equation describing this exponential dilution in Figure 2 is:

$$(CPM)_t = (CPM)_o e^{-\frac{Ut}{P}} \qquad (9)$$

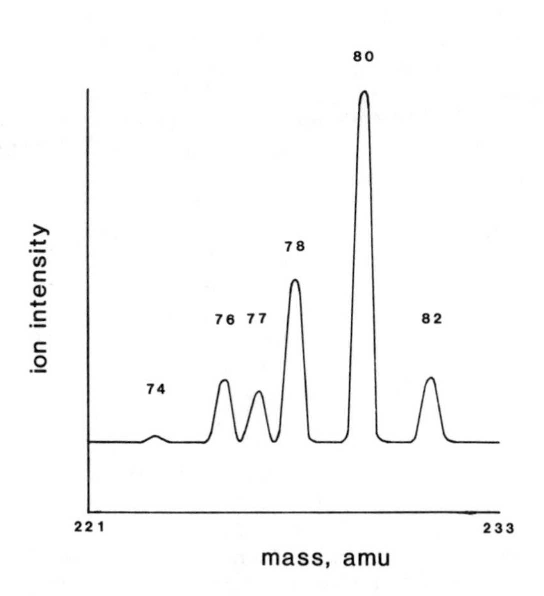

Figure 1. Mass spectrum of Natural (unenriched) Se-NPD$^+$. (Reproduced with permission from Ref. 19. Copyright 1983, American Institute of Nutrition.)

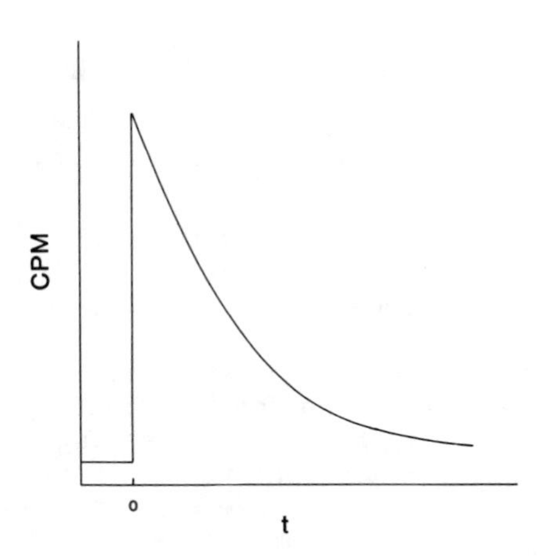

Figure 2. Illustration of exponential dilution from an exponential dilution flask.

where $(CPM)_t$ is the counts per minute at time t; $(CPM)_o$ is the counts per minute at time 0; U is the flow rate of water into the container; and P is the size of the container.

We can extend this concept to tracer studies in humans as well. Let us take as an example the metabolic study of Swanson et al. (20). In this study, 3 groups of women (non-pregnant, early-pregnant and late-pregnant) were placed on a controlled diet containing 150 ug Se/day. On day 8, they also ingested 150 ug Se, but 110 ug of it was natural (unenriched) selenium and 40 ug of it was enriched ^{76}Se incorporated into egg products (7). So we now have the situation illustrated in Figure 3. Here we have a situation analogous to our exponential dilution flask. The equation describing this situation would be:

$$(Se)_t = (Se)_o e^{-\frac{Ut}{P}} \tag{10}$$

where: $(Se)_t$ is the ug of ^{76}Se at time t; $(Se)_o$ is the ug of ^{76}Se at time 0; U is the ug Se flowing in (e.g., 150 ug/day); and P would be the "pool size" (in ug).

In this situation, the most convenient time unit would be days. If we look at, say, urinary excretion (24-hour collections) using days as our time unit, then $(Se)_t$ would be the ug of ^{76}Se in each 24-hour urine, $(Se)_o$ would be the ug of ^{76}Se for the day that the ^{76}Se was ingested; U would be 150 ug Se/day; and P would be their "pool size" in ug.

Plotting 24-hour urinary ^{76}Se excretion for these subjects versus days yields curves very similar to Figure 2. However, there is a more informative way to look at these data. If we take the natural logarithm of both sides of Equation 10 we get:

$$\ln(Se)_t = \ln(Se)_o - \frac{Ut}{P} \tag{11}$$

Thus, a plot of $\ln(Se)_t$ (i.e., ln of the daily ^{76}Se urinary excretion) versus t (i.e., days) should yield a straight line with a slope of -U/P. Since U is known in this case, P can be calculated.

This is illustrated for the non-pregnant subjects in this example in Figure 4. The slope for the first 3 days corresponds to a "pool size" of 230 ug Se. An interesting coincidence is that the plasma selenium concentrations and the estimated plasma volumes (very similar in these subjects) correspond to about 230 ug of circulating plasma selenium. This may, of course, be coincidence, but since plasma and urine can exchange things in the kidneys one might speculate that there is perhaps more to it than coincidence.

The less-negative slope seen for days 11 and 12 indicates a larger additional pool with possibly a slower turnover rate than that of days 8-10.

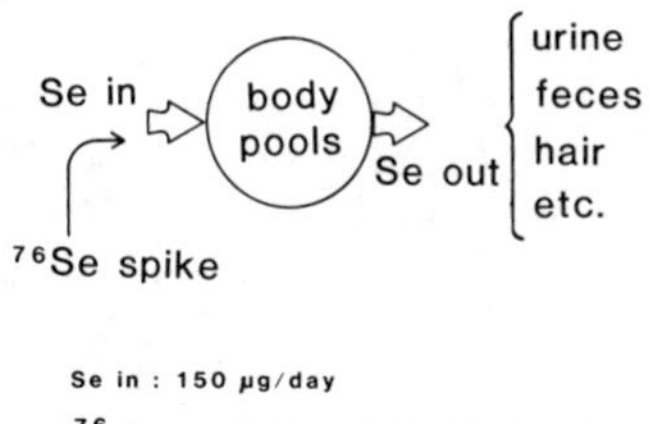

Figure 3. Diagram of the metabolic tracer study of Swanson et al. (20).

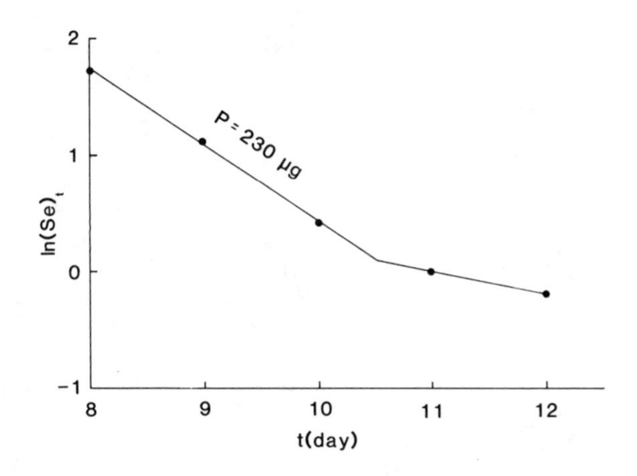

Figure 4. Plot of $\ln(Se)_t$ vs. t for the non-pregnant subjects of Swanson et al. (20). Similar plots for their early-pregnant and late-pregnant subjects gave increasing values of P.

A similar plot for the late-pregnant group of subjects gave a larger P value. However, the difference can not be attributed solely to the fetus, since all three groups of subjects were in slightly positive selenium balance.

Conclusions

The method described herein provides one with an excellent means of determining the selenium content of biological samples and of measuring enriched stable isotopes of selenium in metabolic tracer studies. The sample preparation method is rapid and avoids the problems associated with $HClO_4$. The chelation with NPD combined with the separation capabilities of gas chromatography and the mass discrimination of the mass spectrometer result in an extremely specific method with virtually no chance of interference by other elements in the sample. The method has proven valuable in endogenous food labeling and human metabolic studies with enriched stable isotopes of selenium. Its potential in assessing various selenium body pools is currently being further explored.

Literature Cited

1. Subcommittee on Selenium, National Research Council, 1983, "Selenium in Nutrition, revised edition", National Academy of Sciences, Washington, D.C.
2. Rotruck, J.T.; Pope, A.L.; Ganther, H.E.; Swanson, A.B.; Hafeman, D.G.; Hoekstra, W.G. Science 1973, 179, 588.
3. Keshan Disease Research Group of the Chinese Academy of Medical Sciences. Chin. Med. J. 1979, 92, 471-482.
4. van Rij, A.M.; Thomson, C.D.; McKenzie, J.M.; Robinson, M.F. Am. J. Clin. Nutr. 1979, 32, 2076.
5. Johnson, R.A.; Baker, S.S.; Fallon, J.T.; Maynard, E.P.; Ruskin, J.N.; Wen, Z.; Ge, K.; Cohen, H.J. New England J. Med. 1981, 304, 1210.
6. Janghorbani, M.; Christensen, M.J.K.; Steinke, F.H.; Young, V.R. J. Nutr. 1981, 111, 817.
7. Swanson, C.A.; Reamer, D.C.; Veillon, C.; Levander, O.A. J. Nutr. 1983, 113, 793.
8. Frew, N.M.; Leary, J.J.; Isenhour, T.L. Anal. Chem. 1972, 44, 665;
9. Janghorbani, M.; Young, V.R. Am. J. Clin. Nutr. 1980, 33, 2021.
10. Janghorbani, M.; Young, V.R.; Gramlich, J.W.; Machlan, L.A. Clin. Chim. Acta, 1981, 114, 163.
11. Janghorbani, M.; Christensen, M.J.; Nahapetian, A.; Young, V.R. Am. J. Clin. Nutr. 1982, 35, 647.
12. Carni, J.J.; James, W.D., Koirtyohann, S.R.; and Morris, E.R.; Anal. Chem. 1980, 52, 216.

13. Harvey, B.R. Anal. Chem. 1978, 50, 1866.
14. Turnlund, J.R.; Michel, M.C.; Keyes, W.R.; King, J.C.; Margen, S. Am. J. Clin. Nutr. 1982, 35, 1033.
15. Yergey, A.L.; Vieira, N.E.; Hansen, J.W. Anal. Chem. 1980, 52, 1811.
16. Schwartz, R.; Giesecke, C.C. Clin. Chim. Acta, 1979, 97, 1.
17. Hachey, D.L.; Blais, J.C.; Klein, P.D. Anal. Chem. 1980, 52, 1131.
18. Johnson, P.E. J. Nutr. 1982, 112, 1414.
19. Reamer, D.C.; Veillon, C. J. Nutr. 1983, 113, 786.
20. Swanson, C.A.; King, J.C.; Levander, O.A.; Reamer, D.C.; Veillon, C. Am. J. Clin, Nutr. 1983, 38, 169.

RECEIVED January 31, 1984

8

Iron Absorption in Young Women

Estimated Using Enriched Stable Iron Isotopes and Mass Spectrometric Analysis of a Volatile Iron Chelate

N. S. SHAW, D. D. MILLER, and M. GILBERT—Cornell University, Department of Food Science, Ithaca, NY 14853

D. A. ROE—Cornell University, Division of Nutritional Science, Ithaca, NY 14853

D. R. VAN CAMPEN—U.S. Department of Agriculture, Agricultural Research Service, U.S. Plant Soil and Nutrition Laboratory, Cornell University, Ithaca, NY 14853

A mass spectrometric method was developed for quantifying stable iron isotope tracers present in blood and fecal samples. Volatile Iron acetylacetonate $[Fe(C_5H_7O_2)_3]$ was prepared. A conventional mass spectrometer was used to measure ion abundance ratios of the diligand fragments $[Fe(C_5H_7O_2)_2]^+$ which were formed during electron-impact ionization. Sample isotopic enrichment levels were obtained from standard curves that related ion abundance ratios to enrichment levels. Tracer concentration was calculated from the values for total iron content and enrichment level. The relative standard deviation for the ion abundance measurement was less than 2%. Recovery of tracers from spiked fecal samples ranged from 90% to 104%. The method was used to analyze samples collected from a human study. Iron availability from breakfast meals was determined in 6 young women by giving 7 mg of ^{54}Fe in apple juice on one day and 7 mg of ^{57}Fe in orange juice on the next. Absorption estimated with a fecal monitoring method ranged from -4.8% to 36.5% for ^{54}Fe and from 5.7% to 42.3% for ^{57}Fe. Enrichment of the hemoglobin iron pool by giving 7 mg/day of ^{54}Fe for 7 days was below calculated detection limits for accurate quantification.

Absorption of iron from the diet is an inefficient process which may be enhanced or inhibited by the iron status of the individual consuming the diet, the form of iron in individual foods, and interactions between foods consumed in a single meal (1-4). Because of this, estimates of iron bioavailability obtained from iron absorption measurements are necessary in

0097-6156/84/0258-0105$06.25/0

order to assess the iron adequacy of diets. The accuracy and ease of iron absorption measurements in human subjects can be improved by labeling the dietary iron with radioactive or stable isotope tracers.

To determine iron absorption with tracers, a known amount of an iron tracer is given orally either by itself or with a meal. The appearance of the tracer, either in the feces or in whole blood, is monitored. In the fecal monitoring method, the feces are collected quantitatively. The collection period may last from as few as 3 (5) to as long as 14 days (6). Iron absorption is then estimated from the difference between the total amounts of tracer ingested and excreted. Probably, the most common and serious error in this method is introduced by incomplete fecal collection.

In the hemoglobin incorporation method, absorption is estimated from the amount of tracer appearing in the blood. The principle of this method is based on the observation that absorbed iron is incorporated into circulating hemoglobin 10 to 14 days after dosing (7). A factor of 75% to 90% is usually introduced to correct for the incomplete utilization of absorbed iron for hemoglobin synthesis in normal subjects (8).

The so called extrinsic labeling technique has been used extensively to label foods and meals for iron absorption studies. An inorganic radioiron salt and biosynthetically radioiron-labeled hemoglobin are mixed with food to label the nonheme and heme iron pools in the food respectively. It has been confirmed that absorption of extrinsic tracers is similar to the absorption of intrinsic iron in the food (9, 10).

Stable Isotope Techniques. Although the availability of radioiron isotopes has facilitated our understanding of iron nutrition, their utilization is becoming restricted for safety and ethical reasons, especially when infants, children, and women are involved . The availability of enriched stable iron isotopes (Table I) and methodologies for quantifying them make stable isotopes a feasible alternative to radioisotopes as biological tracers. Neutron activation analysis and mass spectrometry are currently available to nutritionists for quantifying stable isotopes of minerals.

Table I. Isotopic distribution of natural iron and selected lots of enriched ^{54}Fe, ^{57}Fe, and ^{58}Fe*.

Isotope	Abundance in atomic percent: Natural Iron	^{54}Fe-Enriched	^{57}Fe-Enriched	^{58}Fe-Enriched
^{54}Fe	5.82	97.23	0.32	1.14
^{56}Fe	91.68	2.68	13.52	23.74
^{57}Fe	2.17	0.06	86.06	1.86
^{58}Fe	0.33	0.03	0.10	73.26

*Values for enriched isotopes were provided by Oak Ridge National Laboratory (Oak Ridge, TN).

Neutron activation analysis has been applied in studies of plasma clearance of ^{58}Fe (11), iron utilization in pregnant women including maternal-fetal iron transport (12), fecal excretion patterns of orally dosed ^{58}Fe (5), and the effect of oral contraceptives on iron absorption (13). Unfortunately, only ^{58}Fe can be effectively analyzed by neutron activation analysis.

Mass spectrometry is the classic method for ion abundance and isotopic ratio measurements. Ions produced from atoms or molecules introduced into the source of a mass spectrometer are sorted according to their mass-to-charge (m/e) ratios and quantified. Since ions must be in the gas phase in order to be sorted and quantified, elements and compounds of low volatility present special problems. Analysis of minerals requires either the use of specialized instruments or derivitization to volatile compounds.

In thermal ionization mass spectrometry (TI-MS), solid, inorganic compounds may be volatilized from a heated surface. TI-MS is the most precise method for the measurement of isotopic ratios of minerals and has been used to analyze ^{58}Fe in fecal samples collected from a human study (14). The major drawbacks of this technique are the costly instrument and the slow sample through-put. Conventional mass spectrometry produces ions by electron bombardment of the vapor of volatile compounds. This is called electron-impact ionization mass spectrometry (EI-MS). Analysis of iron by EI-MS requires derivitization to volatile forms before introduction into the mass spectrometer. A method has been developed for the synthesis of volatile iron-acetylacetone chelates from iron in blood serum (15). A tetraphenylporphyrin chelate has also been synthesized and used in an absorption study in which ^{54}Fe and ^{57}Fe were given orally (16).

The aim of this study was to further explore the potential and limitations of using stable iron isotopes as tracers and EI-MS in absorption studies. Procedures were developed for preparing iron acetylacetonate from both blood and fecal samples for mass spectrometric analysis. The precision and accuracy of ion abundance measurements were evaluated. In vivo use of stable iron isotope tracers was tested with a human study in which ^{54}Fe and ^{57}Fe were given orally and absorption was estimated with the fecal monitoring and hemoglobin incorporation methods.

Theoretical Considerations

When an enriched isotope is mixed with a sample containing natural iron, the abundance of that isotope in the sample will increase and the change in abundance can be measured with a mass spectrometer. Since the abundance is only a ratio, total iron content of the sample must be determined in order to calculate the amount of the added isotope. Therefore, for quantification of enriched isotope tracers, two measurements must be made on each sample: total iron content and ion abundance ratios.

The enrichment level of an isotope may be expressed as the percent excess of the isotope over its naturally occurring amount:

$$E = (Fe^*/({}_nFe \times A) - 1) \times 100 \qquad (1)$$

where E is the enrichment level in % excess, Fe* is moles of the enriched isotope in a sample, ${}_nFe$ is moles of natural iron in a sample, and A is the fractional abundance on a mole basis of the enriched isotope in the natural iron.

In this study, iron was derivitized with acetylacetone ($C_5H_8O_2$). A typical mass spectrum of iron acetylacetonate [$Fe(C_5H_7O_2)_3$] is shown in Figure 1. The most abundant ion species are the iron-containing, diligand fragments [$Fe(C_5H_7O_2)_2]^+$. The masses representing ^{54}Fe, ^{56}Fe, ^{57}Fe, and ^{58}Fe are 252, 254, 255, 256 respectively. The peak at m/e 254 is the most intense peak and is, therefore, the base peak. The intensities of other ion species are normalized to the value of the base peak and expressed as abundance ratios.

Because of the occurrence of ^2H, ^{13}C, and ^{18}O, and because mass spectrometry sorts ions according to m/e values, the peak intensities at m/e 252-256 do not represent the true proportion of iron isotopes. Table II is an example of how the abundances of the diligand ions deviate from the abundances of iron isotopes. Theoretically, the abundances of the diligand species can be calculated from the abundances of iron isotopes (15). Calculated values may be used as a reference for the accuracy of the experimental values.

Development of Analytical Procedures

Isotopes. Certified iron standards in dilute HCl solution (Fisher Scientific Co.) were used as a source of natural iron. Enriched stable iron isotopes (^{54}Fe and ^{57}Fe) in oxide form were obtained from Oak Ridge National Laboratory. Stock solutions of enriched ^{54}Fe and ^{57}Fe were prepared by dissolving separately 147.5 mg of $^{54}Fe_2O_3$ (purity 97.08%) and 165.3 mg of $^{57}Fe_2O_3$ (purity 86.06%) in 1 ml of concentrated HCl and bringing the final volume up to 100 ml with iron-free distilled water. The isotope concentration was 0.990 mg ^{54}Fe/ml for the ^{54}Fe solution and 1.003 mg ^{57}Fe/ml for the ^{57}Fe solution. A series of ^{54}Fe- and ^{57}Fe-enriched standards were prepared separately by mixing known amounts of the enriched stock solutions with known amounts of natural iron, and the enrichment levels were calculated to be 8.79%-1709% excess for ^{54}Fe standards and 44.79%-2450% excess for ^{57}Fe standards.

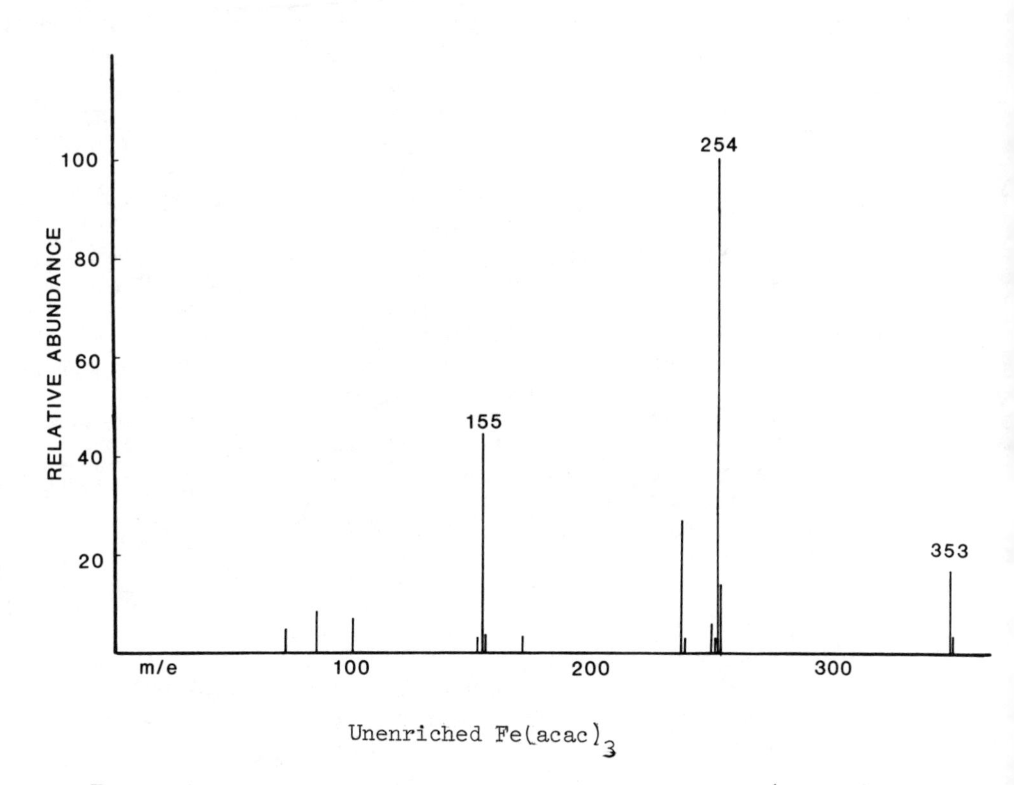

Figure 1. Mass spectrum of iron acetylacetonate $Fe(C_5H_7O_2)_3$ prepared from a unenriched iron standard. Spectra obtained for unenriched blood and fecal samples showed a similar pattern.

Table II. Abundances of iron isotopes and diligand fragments $[Fe(C_5H_7O_2)_2]^+$

	Abundance in atomic percent					
	Natural iron		100% excess ^{54}Fe		100% excess ^{57}Fe	
Isotope	Isotope	Chelate Fragment	Isotope	Chelate Fragment	Isotope	Chelate Fragment
^{54}Fe	5.82	5.16	11.00	9.75	5.70	5.05
^{56}Fe	91.68	81.30	86.64	76.90	89.73	79.57
^{57}Fe	2.17	11.04	2.05	10.43	4.25	12.68
^{58}Fe	0.33	1.62	0.31	1.53	0.32	1.79

Sample Treatments. Blood (1 ml) and fecal samples (1 g dry matter) were ashed on hot plates by sequential treatment with concentrated nitric acid and 30% hydrogen peroxide. The white residue of each sample was dissolved in 3-5 ml of 6 N HCl, and the final volume was brought up to 25 ml with 6 N HCl. Several 0.1 ml aliquots were transferred to test tubes, and iron concentrations were determined by a colorimetric method using the Batho-reagent (17) which contains hydroxylamine hydrochloride (10%), sodium acetate (1.5 M), and bathophenanthroline disulfonate (0.5 mM). The analytical precision of iron quantification was evaluated by measuring the iron concentrations of 13 replicates of one unenriched fecal sample. The mean of these measurements was 365.7 ug per gram of dry feces, with a relative standard deviation of 2.48%.

Anion exchange chromatography was used to separate iron from other cations. The remainder of the acid solutions were added to mini-columns (4 ml Pasteur pipettes, Fisher Scientific Co.) containing anion-exchange resin (Bio-Rad AG 1-X8). In 6 N HCl, iron is anionic ($FeCl_4^-$) and binds to the resin. After washing with 25 ml of 6 N HCl to remove cations, iron was eluted as a cation with 0.5 N HCl.

Synthesis of Acetylacetonate. Acetylacetone (Eastman Kodak Co.) was redistilled and a 0.1 M stock solution in reagent grade chloroform was prepared. All glassware was acid-soaked and thoroughly rinsed with distilled water. Volatile iron acetylacetonate was prepared from iron in the anion-exchange-column eluates or in the standards by first adjusting the pH to 3.5 with 1 M sodium acetate, then shaking the solution in a separatory funnel with 5 ml of the 0.1 M acetylacetone solution, and finally, draining the chloroform phase into a test tube. The color of the complex solution varied from pale yellow to reddish brown, depending on the iron concentration. For mass spectrometric analysis, the chelate solutions were diluted so that the iron concentrations were in the range of 12-20 ug/ml. A 2 ul aliquot of the diluted solution was transferred to a capillary tube, and the solvent was evaporated under mild vacuum before analysis.

Instrumentation. Isotopic measurements were carried out with a DP-102 magnetic sector mass spectrometer (DuPont Instruments) operated in the electron-impact ionization mode. Because the chelate was thermally unstable and decomposed on a GC column, direct probe introduction was used: a capillary tube containing the chelate was placed in the sample cup at the end of the probe. The temperature of the ion-source chamber was set at 140 C and the probe heater was turned off.

Mass spectra were obtained by scanning a mass range of 50-500 a.m.u.. The mass spectra of $Fe(C_5H_7O_2)_3$ prepared from unenriched blood and fecal samples were similar to that of a unenriched natural iron standard (Figure 1), indicating that interfering materials were eliminated by the clean-up procedure. Ion abundances were measured by scanning a mass range of 250-260 a.m.u. at 10 a.m.u./sec. Retention time for each sample ranged from 1.5 min to 4 min. A data reduction program (MC) in the DP-102 software was used to quantify the heights and areas of the peaks of m/e 252, 254, 255, and 256.

Sample Size. Because the accuracy of the intensity measurements depends on sample size, the effect of sample size was examined by measuring the abundance ratios of diligand ions in a series of unenriched iron solutions with concentrations ranging from 6 to 50 ug/ml. The ion abundance ratios for m/e 252, 255, and 256 to m/e 254 were abnormally low when the base peak (m/e 254) intensity was weak; an example is shown in Figure 2. The ratios reached a plateau only when the base peak reached sufficiently high counts: 1.5 million, 1.0 million, and 3.0 million for measurement of 252, 255 and 256, respectively. Data for 252 and 255 measurements were thus accepted only when the height counts of the base peak were greater than 1.5 million (Figure 2). Data obtained from 20 replicates showed that the height ratios were measured with better precision (%S.D. 0.6%-3.2%) than area ratios (%S.D. 2.4%-6.7%) (Table III), peak heights instead of peak areas were, therefore, used in data collection and quantitative analysis.

Memory Effect. A series of ^{54}Fe- and ^{57}Fe-enriched standards was analyzed. Abundance measurements were carried out in order of increasing enrichment with five to seven replicates for each enrichment level. A significant memory effect was observed (Figure 3), which was suspected to result from carryover from a previous sample of a different enrichment level (18). Since

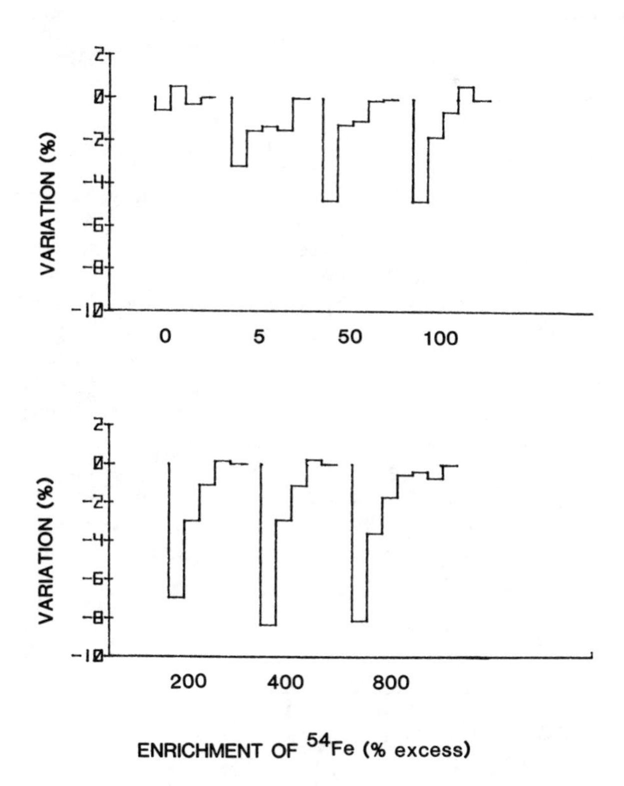

Figure 3. Memory effect for the ion abundance measurement of m/e 252, expressed as a percentage deviation of the measured ratios from that of the last replicate for each enrichment level when enriched standards were analyzed from low to high enrichments.

Table III. Ion abundance ratios of natural Fe $(C_5H_7O_2)_2^+$ values were calculated from measured peak heights or peak areas.

	Ion abundance ratio			
	Height		Area	
Ion species	Ratio	% S.D.	Ratio	% S.D.
252/254	6.97	1.3	6.68	5.0
255/254	14.77	0.6	14.69	2.4
256/254	1.21	3.2	1.94	6.7

the carryover effect would have least influence on the last replicate of a given enrichment level, the magnitude of the memory effect was expressed as a percentage deviation of the measured ratio of each replicate from that of the last replicate. As shown in Figure 3, the first three measurements were seriously contaminated by the previous sample of lower enrichment, and the memory effect was especially significant as enrichment levels increased. However, the measured ratios stabilized after three to four replicates of a particular sample were run. Based on these observations, two steps were taken to reduce the memory effect: first, preliminary measurements were carried out on samples to estimate the enrichment levels, then the samples were analyzed along with the enriched standards in sequence from low to high enrichments; second, six replicates were measured for each sample, and only the last three measurements were accepted.

Standard curves. Standard curves for ^{54}Fe and ^{57}Fe were established by plotting the measured ion abundance ratios of the corresponding diligand ions against the enrichment level of each isotope in the series of enriched standards. Linear curves were obtained for both isotopes (Figure 4), and the measured ratios were highly correlated with the theoretical values ($r > 0.9998$) by linear regression analysis. Relative standard deviations, used as an expression of analytical precision, are listed in Table IV. With values for the precision of these measurements and for the slope of the standard curves, the detection limit was calculated according to Skogerboe et al. (19) to be 7.3% excess for ^{54}Fe and 19.4% excess for ^{57}Fe.

Analysis of Spiked Fecal Samples. To simulate fecal samples taken from subjects receiving oral doses of stable isotopes, ^{54}Fe- and ^{57}Fe-spiked fecal samples were prepared by adding known amounts of the enriched stock solutions of each isotope to aliquots of a homogenized, unenriched, pooled fecal sample. After ashing, the iron concentrations of these samples were determined by the Batho-reagent method. Iron acetylacetonate was then prepared and isotopic measurements of the samples were carried out along with the enriched standards. The amount

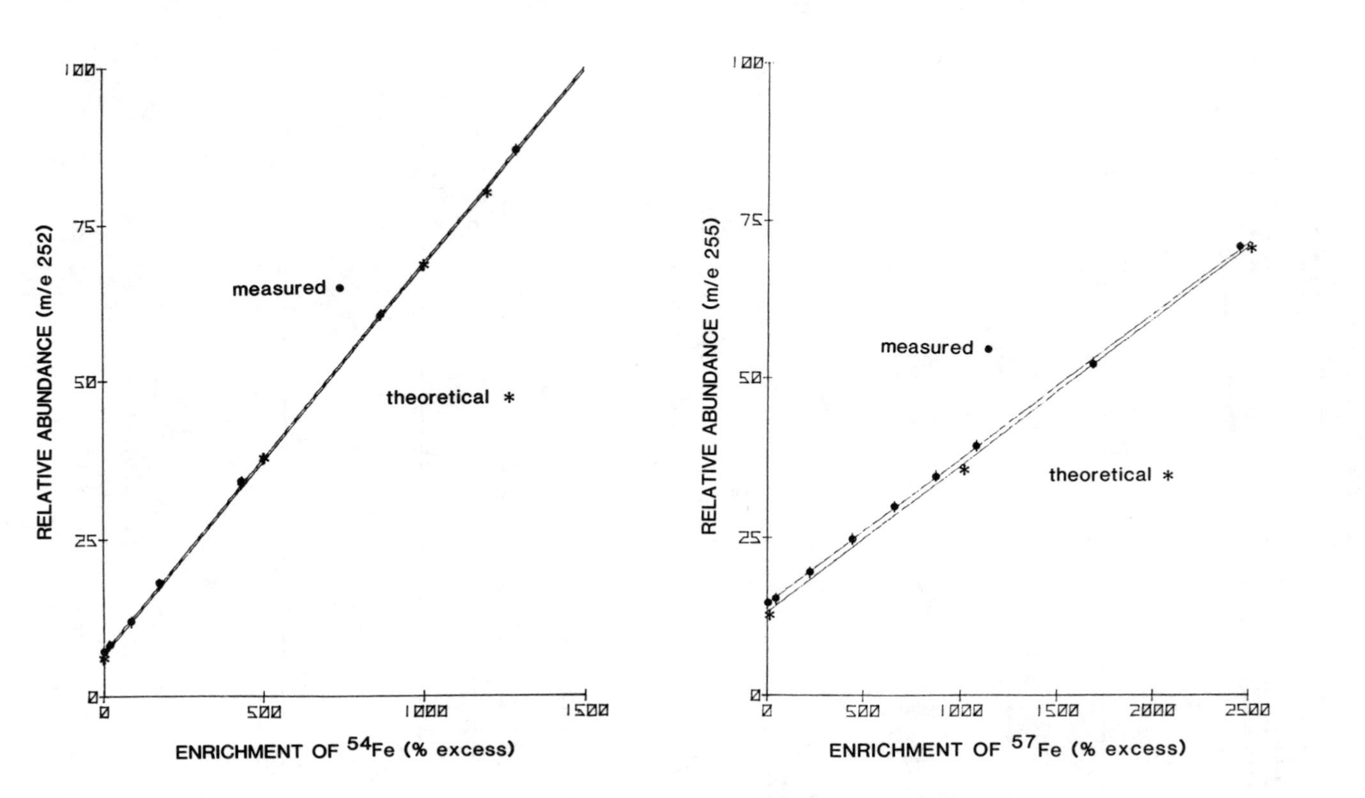

Figure 4. Standard curves for the ion abundance ratio measurements of ^{54}Fe and ^{57}Fe.

Table IV. Precision of abundance measurements.

^{54}Fe standards			^{57}Fe standards		
Enrichment (% excess)	252/254 ratio	Precision % S.D.	Enrichment (% excess)	255/254 ratio	Precision % S.D.
0	7.22	1.1	0	14.88	1.3
88	12.67	0.5	45	16.06	1.7
175	18.26	0.8	223	19.72	0.8
433	34.12	0.8	442	24.84	0.4
867	59.33	0.5	658	30.28	0.8
1291	85.18	0.2	1685	53.72	0.1

of ^{54}Fe or ^{57}Fe recovered from each sample was calculated from the total iron content and the measured enrichments as follows:

$$Fe^* = {}^{t}Fe - {}^{n}Fe \qquad (2)$$

$$E = 100 \times Fe^* / ({}^{n}Fe \times W) \qquad (3)$$

Substituting (3) into (2) and rearranging yields

$$Fe^* = ({}^{t}Fe \times E \times W) / (100 + E \times W) \qquad (4)$$

where Fe* is mg of added tracer in the sample, ^{t}Fe is mg of total iron, ^{n}Fe is mg of natural iron, E is the enrichment level of the tracer, and W is a constant representing the natural abundance of the added isotope in weight fraction (0.0562 for ^{54}Fe and 0.0222 for ^{57}Fe). Recoveries of the spiked isotopes were: 93% to 102% for ^{54}Fe and 91% to 104% for ^{57}Fe (Table V).

Table V. Recovery of 54Fe and 57Fe from spiked fecal samples.

Recovery of 54Fe spikes			Recovery of 57Fe spikes		
Total Fe (ug)	Added 54Fe (ug)	Recovery (%)	Total Fe (ug)	Added 57Fe (ug)	Recovery (%)
700	25.8	93	755	30.1	100
845	24.8	101	740	30.1	91
860	49.5	100	885	100.0	104
797	99.0	98	1107	301.0	98
820	99.0	102	1058	301.0	98

Human Study

A study for examining the feasibility of using stable iron isotopes to measure iron absorption by human subjects was carried out as part of a 12 week study of the effects of exercise on riboflavin requirements (20). The study was carried out in the Francis Johnson-Charlotte Young Human Nutrition Unit at Cornell University. The protocol for the study was approved by the Cornell Committee on Human Research Subjects.

The method of the selection of subjects has been reported in detail (20). Twelve healthy, female university students and staff aged 19 to 27 years participated. Their average weight and height were 60.0 kg (± 4.1 kg) and 163.1 cm (± 4.8 cm), respectively. They agreed to participate in the study after being informed of the purpose of the research and its potential hazards. A single basic daily menu was used throughout the study period for all subjects. Details of meal preparation have been reported (20). The diet provided 11 mg of iron daily. A supplement of 10 mg of iron as ferrous sulfate was given daily with orange juice at breakfast. All the subjects took three meals daily in the metabolic unit. On selected days, about 7 mg of the iron supplement were replaced by enriched stable iron isotopes (Table VI). Polyethylene glycol (PEG) was added as a fecal marker. Hematological parameters (plasma iron, total iron binding capacity, free erythrocyte protoporphyrin, serum ferritin) were measured as indices of iron status. The values for these parameters were within normal ranges for all the subjects except two (21). The two subjects had serum ferritin values lower than 10 ng/ml. Iron absorption was estimated by the fecal monitoring and hemoglobin incorporation methods.

Table VI. Experimental design of the iron absorption study.

Day	Dosing/Sampling
	Fecal monitoring method
1 or 2*	6.73 mg ^{54}Fe in apple juice
1 or 2*	6.94 mg ^{57}Fe in orange juice
1 - 8	fecal collections
	Hemoglobin incorporation method
9 - 15	6.73 mg ^{54}Fe in orange juice
29	blood sample drawn

*On day 1, half of the subjects received ^{54}Fe in orange juice and half received ^{57}Fe in apple juice. On day two, subjects were given the isotope they did not receive on day 1. Two grams of PEG were given with the first dose.

<u>Fecal Monitoring Method</u>. The subjects were instructed to bring their stool samples back to the metabolic unit as soon as was possible after collection. Each sample was carefully labelled with time, date, and subject identification, and was placed in a freezer in the metabolic unit. The samples were freeze-dried, and allowed to equilibrate for 24 hours under ambient conditions. The dry weights of the samples were taken, and the samples were powdered and mixed thoroughly to provide homogeneous portions for analysis.

The fecal marker (PEG) was analyzed by the turbidimetric method of Malaware and Powell (22). Total recoveries of ingested PEG in feces ranged from 20% to 99.7%. Three subjects excreted less than 60% of the ingested PEG, whereas the rest had an average recovery of 85.3% (range 73%-99.7%).

Iron concentration in each fecal sample was measured after ashing and total iron content was calculated from the iron concentration and dry weight of the sample. Iron acetyl-acetonate was prepared and abundance measurements of samples were carried out along with the enriched standards. Since most of the fecal samples were labeled with both ^{54}Fe and ^{57}Fe, the content of each isotope in the sample were calculated as follows:

$$^{t}Fe = {^{n}Fe} + {^{54}Fe^*} + {^{57}Fe^*} \qquad (5)$$

$$^{54}Fe^* = {^{54}E} \times 0.0562 \times {^{n}Fe} / 100 \qquad (6)$$

$$^{57}Fe^* = {^{57}E} \times 0.0222 \times {^{n}Fe} / 100 \qquad (7)$$

After substituting equation 6 and 7 into 5 and rearranging, the amount of natural iron can be obtained given the total iron content and the enrichment levels:

$$^{n}Fe = {^{t}F} / (1 + 0.000562\ {^{54}E} + 0.000222\ {^{57}E}) \qquad (8)$$

The content of each isotope tracer in a sample was then calculated from values for natural iron content and enrichment level (Equation 6 and 7). A computer program was set up to carry out all these calculations.

Some examples of the excretion patterns of both PEG and ^{54}Fe are shown in Figure 5. Based on PEG recovery and regularity of bowel movement, only six subjects were selected for analysis. PEG appears to be a good qualitative marker of iron tracer excretion. However, since PEG recovery was lower than ^{54}Fe recovery for two out of six subjects, the reliability of quantitative corrections based on the PEG recovery are questionable. The total amounts of ^{54}Fe or ^{57}Fe tracers excreted by each subject were calculated by summing the amounts in each individual sample. Absorption calculated as the difference between the amounts of the ingested and excreted tracers without corrections for PEG recovery was expressed as a percentage of the tracer doses: for ^{54}Fe in apple juice, it ranged from -4.8% to 36.5%, equivalent to -0.32 to 2.46 mg of tracer per day; for ^{57}Fe in orange juice, it ranged from 5.7% to 42.3%, equivalent to 0.40 to 2.94 mg of tracer per day. Although vitamin C has been shown to enhance iron absorption (23), and in this study, orange juice provided about 100 mg of vitamin C (20), the difference between the two dietary treatments

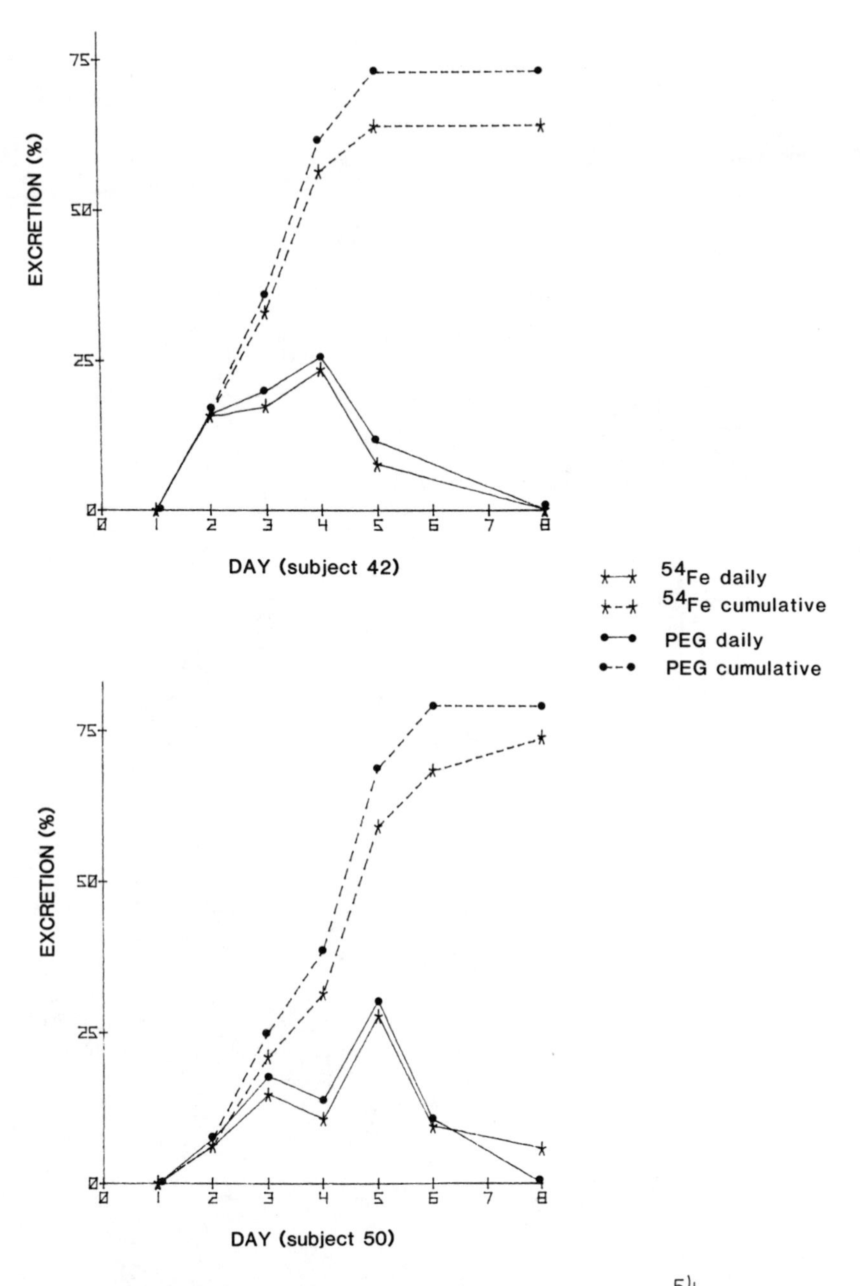

Figure 5. Daily and cumulative excretion of PEG and ^{54}Fe in two subjects. Feces were collected for 8 d after 7 mg of ^{54}Fe and 2 g of PEG fecal marker were served with apple juice. Results for a third subject are on page 120.

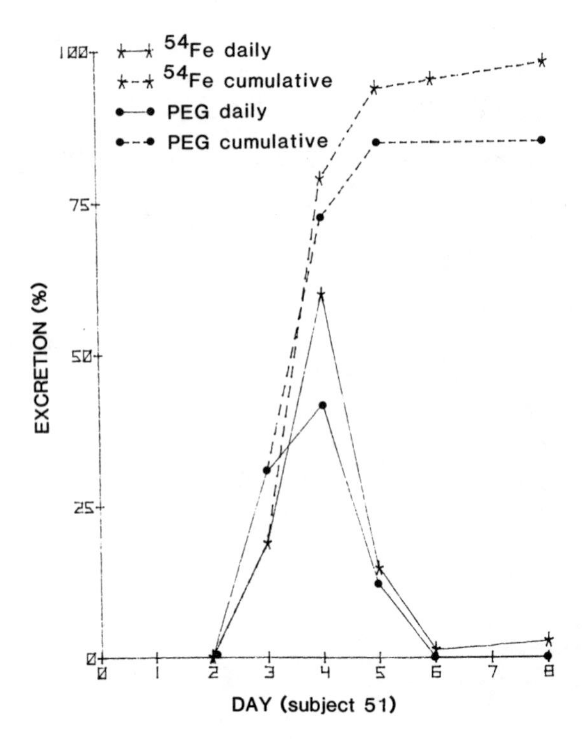

Figure 5. Continued. Daily and cumulative excretion of PEG and ^{54}Fe in a third subject. First two subjects' results appear on page 119.

was not significant when the paired data for each subject were analyzed by Student's t test.

The Hemoglobin Incorporation Method. The hemoglobin concentration of each blood sample was determined, and iron chelates were prepared for abundance measurements. The blood volume was estimated from the equation of Jarnum (24):

$$BV = 0.414 \times H^3 + 0.0328 \times W - 0.030 \quad (9)$$

where BV is blood volume in liters, H is height in meters, and W is body weight in kilograms. Total iron content in blood (^{B}Fe, mg) was calculated from hemoglobin concentration (Hb, mg/l), blood volume (BV), and the iron content of hemoglobin (0.334%):

$$^{B}Fe = Hb \times BV \times 0.00334 \quad (10)$$

The total amount of the ^{54}Fe tracer in blood was calculated from the hemoglobin concentration, blood volume, and enrichment level of ^{54}Fe in blood (^{B}E):

$$^{54}Fe^* = 1.877 \times 10^{-6} \times Hb \times BV \times {}^{B}E \quad (11)$$

Percentage absorption was calculated from the values for total ^{54}Fe in the blood and the total ingested ^{54}Fe for the 7 day period. The range for 12 subjects was 0 to 17.8 %, which was equivalent to 0 to 1.05 mg of ^{54}Fe tracer per day (assuming 85% hemoglobin utilization). Table VII is a comparison of absorption data of iron in orange juice obtained with the fecal and hemoglobin monitoring methods.

Table VII. Absorption of stable iron isotopes in orange juice measured with the fecal monitoring and hemoglobin incorporation methods

	Fecal monitoring		Hemoglobin Incorporation	
Subject	Absorption with PEG correction (%)	Absorption w/o PEG correction (%)	Enrichment of hemoglobin pool (% excess)	Absorption (%)
41	-*	10.6	2.29	5.5
42	21.0	42.3	1.57	3.8
46	20.1	20.3	4.11	7.8
50	6.0	25.3	1.26	3.2
51	-*	5.7	3.23	9.0
52	11.7	17.8	7.78	15.6

*Recovery of isotope tracers was higher than PEG recovery.

Discussion

Variability in Absorption Estimates. In this study, the occurrence of a negative absorption value for one subject and the absence of a significant vitamin C effect raise some questions about the accuracy of the method. However, the expected changes in absorption due to dietary treatments may be masked by the analytical variations associated with absorption measurements and biological variabilities of iron absorption. Analytical variations can be introduced at several stages of the analytical procedures: incomplete fecal collection, inhomogeneous samples, iron contamination, incomplete colorimetric reaction, non-quantitative recovery after chemical ashing, and variations in isotopic measurements due to ion statistics, memory effects, instrument drift, etc. Some of these are not as serious as others, for example, contamination with natural iron would not affect the estimate of tracer concentrations provided it occurs before the total iron content is measured.

Errors associated with iron quantification and abundance measurements are common to both fecal monitoring and hemoglobin incorporation methods. According to Janghorbani and Young (25), acceptable absorption estimates can be obtained if the precision of these measurements is kept below 5%. The analytical procedures developed in this study are thus considered satisfactory since the relative standard deviation was 2.48% for total iron quantification and less than 2% for ion abundance determinations.

Incomplete fecal collection introduces an error unique to the fecal monitoring method. Careful instruction of the subjects regarding collections will help to ensure complete collection, however, without the use of a suitable fecal marker that is completely unabsorbable, has the same mobility as ingested iron, and can be accurately quantified in the stool, absolute certainty of complete collection cannot be achieved. PEG, chromic oxide, and radio-opaque plastic pellets are commonly used as fecal markers. Chromic oxide and plastic pellets are usually considered as markers for the solid phase, whereas PEG marks the liquid phase (26). As shown in this study, PEG seems to have an excretion pattern similar to iron, yet it failed as a quantitative marker. This was probably due to interference in PEG quantification and reexcretion of the absorbed iron tracers (27). Unabsorbable compounds of chromium and barium have also been used as markers in iron and zinc absorption studies (6, 28). Chromium has potential as a marker for stable isotope studies because it has four stable isotopes and can form volatile chelates, thus permitting quantification by EI-MS.

In the hemoglobin incorporation method, estimation of blood volume from height and weight may introduce a 5% error (29). A more serious source of error is the low enrichment level obtainable in whole blood: the hemoglobin iron pool is large (1.5-2g) and daily iron absorption is relatively small (about 1 mg or less). Most of the subjects in this study did not absorb enough of the tracer to enrich the blood pool up to the detection limit. Either lengthening the study period from 7 days to 2 weeks or increasing the tracer dose from 7 mg daily to 10 mg would significantly raise the enrichment level. Because fecal collections are not required with this method, it is an extremely attractive alternative to the fecal monitoring method. Further work to improve the precision of the enrichment determination seems justified since this would reduce the detection limits for the measurement of stable isotope tracers in blood and make the use of the method more feasible.

Biological variabilities include intra-subject and inter-subject variabilities. Intra-subject variability is best demonstrated in the day-to-day variation in absorption when the same test meal labelled with different iron tracers is given to the same subject on two successive days. A threefold to fourfold difference has been reported with standard deviations of -50% to +80% (30) within a group of normal subjects. This variability can be reduced to about 35% by giving iron doses over several days and using the average absorption value instead of the individual values (31). Inter-subject variability results from differences in individual iron status; subjects with low iron stores absorb more dietary iron than those with larger iron stores (2). The significance of these variabilities depends on the purpose of the study. If comparison of dietary treatments is desired, it is desirable to reduce these variabilities as much as possible so that the changes due to treatment will not be masked. If the interest is in iron requirements, then the variabilities must be considered for proper dietary recommendations.

Comparison of Absorption Data. Little correlation was found between the absorption data obtained with the hemoglobin incorporation and fecal monitoring methods with or without correction for PEG recovery (Table VII). There are several possible reasons for this poor correlation.

a) Total or partial loss of stool samples occurred and correction based on PEG recovery was inadequate.

b) A portion of the ingested iron was temporarily retained by the mucosal cells and was excreted in the feces after fecal collection was stopped (6).

c) Intra-subject variability resulting from daily variations of the gastrointestinal secretion and motility makes the fecal monitoring data highly variable. It has been reported that this variability can account for up to half of the overall variability in iron absorption within a group of normal subjects (29).

d) The different experimental designs of the two methods may provide different kinds of information. The fecal method measured absorption from a single dose, while the hemoglobin method provided an average value spread over 7 doses. Since nutrient balance is a dynamic state that fluctuates between negative and positive balance (32), absorption of a single dose may not be representative, and an average value probably is a better indication of actual iron absorption.
The data available from the study described here do not permit a conclusive statement as to which of the above reasons offers the best explanation of the results. However, in future planning of human metabolic studies, the factors mentioned above should be taken into consideration.

Comments and Perspectives

The analytical procedures presented in this study are relatively simple and sample through-put through the mass spectrometer is reasonable (15 min per replicate). The results indicate that the method has promise as an alternative to radioisotope tracer methods. Long-term studies and studies with infants, children, and women are feasible without hazards to the subjects. Although the precision of EI-MS abundance measurements is not comparable to that of TI-MS, the EI mass spectrometer is much more widely available. More stable iron chelates that would permit GC sample introduction and reduce the memory effect problem would significantly improve the method. A potentially serious drawback of stable isotope tracers is that the amounts used are more in the range of substrate than tracer levels. A related question pertains to the validity of the extrinsic labeling technique when stable isotope tracers are used. In an ongoing study in our laboratory, we are attempting to compare the absorption of intrinsic and extrinsic stable iron isotope tracers.

Literature Cited

1. Pirzio-Biroli, G.; Finch, C. A. J. Lab. Clin. Med. 1960, 55, 216.
2. Cook, J. D.; Lipschitz, D. A.; Miles, L. E. M.; Finch, C. A. Am. J. Clin. Nutr. 1974, 27, 681.
3. Young, V. R.; Janghorbani, M. Cereal Chem. 1981, 58, 12.
4. Monsen, E. R.; Hallberg, L.; Layrisse, M.; Hegsted, D. M.; Cook, J. D.; Mertz, W.; Finch, C. A. Am. J. Clin. Nutr. 1978, 31, 134.
5. Janghorbani, M.; Ting, B. T. G.; Young, V. R. J. Nutr. 1980, 110, 2190.
6. Boender, C. A.; Verloop, M. C. Brit. J. Haemat. 1969, 17, 45.
7. Hahn, P. F.; Bale, W. F.; Lawrence, E. O.; Whipple, G. H. J. Exp. Med. 1939, 69, 739.

8. Skarberg, K.; Eng, M.; Huebers, H.; Marsaglia, G.; Finch, C. Proc. Nat. Acad. Sci. (U. S.) 1978, 75, 1559.
9. Layrisse, M.; Martinez-Torres, C. Am. J. Clin. Nutr. 1972, 25, 40.
10. Cook, J. D.; Layrisse, M.; Martinez-Torres, C.; Walker, R.; Monsen, E.; Finch, C. A. J. Clin. Invest. 1972, 51, 805.
11. Lowman, J. T.; Kirvit, W. J. Lab. Clin. Med. 1963, 61, 1042.
12. Dyer, N. C.; Brill, A. B., in "Nuclear Activation Techniques in the Life Sciences"; International Atomic Energy Agency : Vienna, 1972; p.469.
13. King, J. C.; Raynolds W. L.; Margen, S. Am. J. Clin. Nutr. 1978, 31, 1198.
14. Turnlund, J. R.; Michel, M. C.; Keyes, W. R.; King, J. C.; Margen, S. Am. J. Clin. Nutr. 1982, 35, 1033.
15. Miller, D. D.; Van Campen, D. Am. J. Clin. Nutr. 1979, 32, 2354.
16. Johnson, P. E. J. Nutr. 1982, 112, 1414.
17. Smith, G. F. "The iron reagents"; G. F. Smith Chemicals: Columbus, OH, 1963; p.30.
18. Buckley, W. T.; Huckin, S. N.; Budac, J. J.; Elgendorf, G. K. Anal. Chem. 1982, 54, 504.
19. Skogerboe, R. K.; Heykey, A. T.; Morrison, G. H. Anal. Chem. 1966, 38, 1821.
20. Belko, A. Z.; Obarzanek, E.; Kalkwarf, H. J.; Rotter, M. A.; Bogusz, S.; Miller, D. D.; Haas, J. D.; Roe, D. A. Am. J. Clin. Nutr. 1983, 37, 509.
21. Shaw, N. S. Ph.D. Thesis, Cornell University, New York, 1984.
22. Malawere, S. J.; Powell, D. W. Gastroenterology 1967, 53, 250.
23. Cook J. D.; Monsen, E. R. Am. J. Clin. Nutr. 1977, 30, 235.
24. Jarnum, S. Scand. J. Clin. Lab. Invest. 1959, 11, 269.
25. Janghorbani, M.; Young, V. R., in "Advances in Nutrition Research"; Draper, H. H.; Ed.; Plenum Press : New York, 1980; Vol. 3, p. 127.
26. Allen, L. H.; Reynolds, W. L.; Masgen, S. Am. J. Clin. Nutr. 1979, 32, 427.
27. Johnson, P.E. personal contact.
28. Payton, K. B.; Flangen, P. R.; Stinson, E. A.; Chodirker, D. P.; Chambalain, M. J.; Valberg, L. S. Gastroenterology 1982, 83, 1264.
29. Cook, J. D.; Lipschitz, D. A. Clin. Heamat. 1977, 3, 567.
30. Cook, J. D.; Layrisse, M.; Finch, C. A. Blood 1969, 33, 421.
31. Brise, H.; Hallberg, L. Acta Med. Scand. 1962, 3, 7.
32. Hegsted, D. M. Nutr. Rev. 1952, 10, 257.

RECEIVED January 31, 1984

9

Bioactive Trace Metals and Trace Metal Stable Isotopes

Quantitative Analysis Using Mass Spectrometric Techniques

D. B. CHRISTIE, M. HALL, C. M. MOYNIHAN, K. M. HAMBIDGE, and P. V. FENNESSEY

University of Colorado Health Sciences Center, Denver, CO 80262

Understanding the factors controlling the bioavailability of trace metals in human nutrition is central to the treatment of deficiencies of these elements. Nutrition science has been hampered in these studies by techniques that require pharmacological doses (10 to 100 X physiological), radioisotopes or equipment that is expensive and of limited availability. In spite of these limitations, a few laboratories are obtaining data on this important topic. The advent of high resolution gas chromatography has opened the door for the quantitative analysis of trace elements from physiological sources using relatively inexpensive equipment. We will present data on the analysis of trace elements at the ppb and sub-ppb range. In achieving this level of sensitivity, we have encountered and overcome a number of technical problems which will also be discussed. Analysis of the stable isotopes of trace metals has presented itself as a complex and difficult analytical problem. The causes and solutions to a number of the problems encountered in our laboratory will be discussed. The techniques we are currently using involve both standard gas chromatography/mass spectrometry and fast atom bombardment mass spectrometry techniques.

There appears to be an ever increasing momentum in the field of human nutrition to measure true bioavailability of a wide variety of physiologically important metals. A key driving force for these studies is a desire to understand the important mechanistic steps in both the absorption and transport of metals across the

0097–6156/84/0258–0127$06.00/0

gut wall. An understanding of the processes will lead to changes in dietary supplementation and ultimately to healthier individuals.

The ideal trace metal bioavailability study would be carried out at no risk and on a human population. However, many of the early investigations could only be accomplished using radioactive trace metals and were carried out on animals. A few studies have been reported using these radioactive compounds on adult human volunteers and these have provided us with important data for human nutrition.(1-6) Unfortunately, two important groups of patients, namely, pregnant women and infants, cannot be studied using radioactive material. For these groups the most promising data appear to be from the area of stable isotope analysis. This technique is apparently of low risk and can be applied to a wide variety of both healthy and sick individuals. (7-12) Further, stable isotopes are available for many of the metals of interest in human nutrition.

The Achilles Heel in the use of stable isotopes for human nutritional studies is the methodology available for the detection of the enriched stable isotope. Unlike the radioactive studies where one can use a relatively cheap and very accessible technique such as scintillation counting, there appear to be only two methods for the analysis of metal isotopes: mass spectrometry (10-18) and neutron activation analysis.(7-9) To date, both methods have proven to be less than desirable because of the time-intensive nature of the analysis.

Mass spectrometry of zinc isotopes has been realized using either chelates on a solids probe(10,11,17) or thermal ionization of purified solutions(12). Both of these approaches require a chemical separation of all of the metals and this separation must be accomplished in an environment free of contamination from the metal(s) of interest in the part-per-billion range (19). Neutron activation also requires a set of separation steps. In this case the requirement for a contamination-free environment is the same but the chemical separation is mainly to remove sodium and chlorine (7).

In addition to the human labor problems created by chemical separation and ultra-clean environments, both neutron activation and thermal ionization mass spectrometry require unique equipment that is not available to most research groups. Because of these factors progress in trace metal analysis has been slow and practical data acquisition from large populations is essentially non-existent.

Gas chromatography and gas chromatography mass spectrometry offer a potential solution to the problems mentioned above. For example, a separation of a variety of interesting trace metals is achieved by the gas chromatograph and the methodology for derivatization to form a chelate from an HCl solution of dry ashed material has been achieved at high levels of efficiency (20). In addition, the flame ionization detector of a gas chromatograph can be used to generate quantitative measurements of trace elements from physiological samples. The gas chromatographic detector can be replaced by a mass spectrometer and one can obtain measurements on isotopic distributions from all of the trace elements in each sample. Although this approach would appear to be the method of choice, it has been abandoned by most investigators. This is because the precision, or lack of precision, of 2-100% has been reported for gas chromatography/mass spectrometry of trace metal chelates (12). The problems that have lead to this imprecision and their solutions are the subject of this report.

Gas Chromatography

The key to the analysis of a trace metal by gas chromatography is the formation of a volatile chemical complex. In early studies this was accomplished using derivatives of β-diketones (21-23). These compounds formed chelates that were stable in the gas phase and would separate many of the common elements. In our review of the literature we were impressed by the work of two Polish scientists who were studying zinc, copper and nickel in marine sediments(20,24). These investigators were able to quantitate these elements at the part-per-billion level (50 ng/μl injected). These results were obtained using diethyldithiocarbamates as organic chelators. Our goal was to develop a technique that would provide trace metal concentration data on samples obtained from 0.1 to 0.2 ml blood samples. Using the carbamate on an OV-101 column we were able to measure 500 ng samples but at lower levels the data was too variable. After switching to a capillary column and a dry needle injector we were able to reduce our level of detection to less than 1.0 ng. This level is a factor of 20 less than what we would expect to see in a patient sample and is adequate for the quantitative analysis. A typical gas chromatogram from the injection of a complex mixture of 20 ng each of metals is shown in Figure 1. The peaks are sharp and the separation such that one can carry out a low concentration analysis on any one of a number of interesting trace metals.

Quantitative analysis of trace metals has been accomplished using this system and a hydrocarbon as a reference material. The concentration data obtained in this manner had a coefficient of variation of 8% (see Figure 2), and when compared to atomic absorption, data collected from the same samples gave concentrations that were essentially the same. The use of hydrocarbons as a standard is recognized as a problem and we plan to evaluate other metal chelates for this purpose. The internal standard could then be added to our original samples prior to dry ashing.

In interpreting the gas chromatographic data from our initial plasma extracts we were surprised to see a variable size

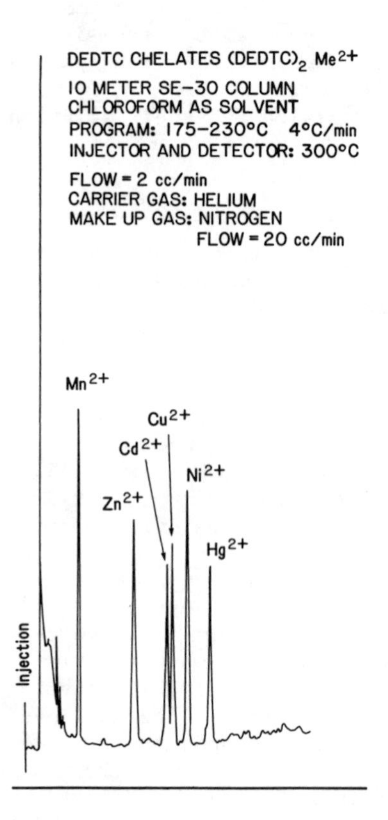

Figure 1. Gas chromatograph trace using F.I.D. of a standard mixture of 20 ng each metal chelates.

in the area of the iron chelate peak. This observation immediately raised the issue of possible poor reproducibility throughout the extraction of all the metals. Additional work indicated that the variability was associated with the pH of the chelation reaction. In order to find the optimum pH, we chelated and extracted a series of metals from the same stock solution at different pH values. The iron was extracted well at pH < 4 while the zinc and copper showed a maximum at pH 6.5. In our present work we use a pH of 6.0 because of our interest in zinc and copper and at this pH one can still see an iron peak. However, if our interest were to change to iron we would certainly move the pH lower. We have concluded that some of the variability that we and others have observed in our analysis could have been accounted for by a lack of careful control of the pH of our solutions.

Mass Spectrometry

The real drive in using gc and gc/ms techniques for the analysis of trace metals is measuring changes in isotopic enrichment. The quantitative data provided by the gas chromatograph is important but we consider it only a spin-off of the real goal of this research. This final goal has also been the source of real frustration. After we had attained good peak shape and separation from the gas chromatograph, we began taking data on a low resolution mass spectrometer. This system had an "all glass" interface and the total ion current showed separate peaks for zinc, copper and nickel standards. However, the mass spectrum of the zinc peak showed contamination from both nickel and copper. This contamination appeared in the zinc peak after changing both the column and injector. Since the zinc peak was separate from the copper and nickel and still contained contamination from copper and nickel, we suspected the interface between the gas chromatograph and the mass spectrometer. After reviewing the manufacturer's design for the interface we found that the end of the gas chromatograph column terminated in a small chamber that was connected to the mass spectrometer source through a fine capillary interface tube. The chamber was connected to the glass line tubing using a standard ¼ inch stainless steel connector. We could envision that the chelated sample would swirl around this connection while waiting to be aspirated into the source. After modification of this "all glass" interface into a unit where the column effluent did not come into contact with metal we were able to eliminate the copper and nickel contamination of the zinc peak.

The next problem that we had to face was the issue of memory. This can best be seen from the results of an experiment where we repeatedly injected a series of eleven zinc chelates from the same container and monitored the ratio of zinc-64 to both zinc-67 and zinc-68 (see Figure 3). This was followed by an equal number of injections where the isotope ratio was enriched and, finally, by a series of injections of the original solution. We expected to see the data follow a square wave pattern but found the rather slumped pattern seen in Figure 3. The ratio did not return to baseline even after eleven injections of the original solution. Initially we thought that this was due to thermal decomposition of the chelate. We were able to obtain some improvement in our ability to return to baseline by controlling the injector and interface temperatures but for the most part the memory remained.

While still facing what we perceived as a thermal problem we began a program to synthesize the bis-trifluoroethyldithiocarbamate. This compound was reported to chromatograph at a lower temperature (circa 100°C less) than the diethyl analog (25). This would allow us to lower our injection and transfer lines by 100°C and reduce possible decomposition problems. We synthesized the trifluoro compound and we were surprised by the experimental data which showed a clear memory effect between the trifluoroethyl and diethyl chelates. This effect was exemplified by peaks of intermediate retention time when the pure trifluoro or pure ethyl chelates were injected in sequence. For example, after injection of the diethyl derivative which elutes at approximately 215°C, we would inject the trifluoro derivative and expect to see a peak at approximately 120°C. In fact the peak would come out at 180°C. A second injection would produce a peak at 150°C and so on until the peak would elute at the expected 120°C. The same type of progressive increase was observed if one then injected the pure diethyl compound after first stabilizing the trifluoro derivative's elution at 120°C.

The observation using the two chelators led us to reevaluate the causes for the memory. We began to consider the possibility of some interaction between the zinc and the gas chromatograph column. A review of data available on zinc diethyldithiocarbamate chelate structure indicated that these compounds had weak coordination with a fifth ligand (26). We began to think that this fifth ligand could interact with either a second zinc atom or with the gc column support material (e.g. oxygen or silicone). A possible method to remove this interaction would be to supply a fifth ligand binding molecule either in the carrier gas or in the chelate solution.

Our first attempt was to try ammonia. This was added to the chelator solution in the form of ammonium acetate and proved to be very effective in reducing the interaction of the chelated zinc with the column. One experiment that exemplifies the effect of ammonium acetate is shown in Figure 4. Here we put a mixture of both the bis-trifluoro and the diethyl chelate on our injector. The first gas chromatogram (gc) was obtained by using a mixture of copper and zinc where ammonium acetate has been added

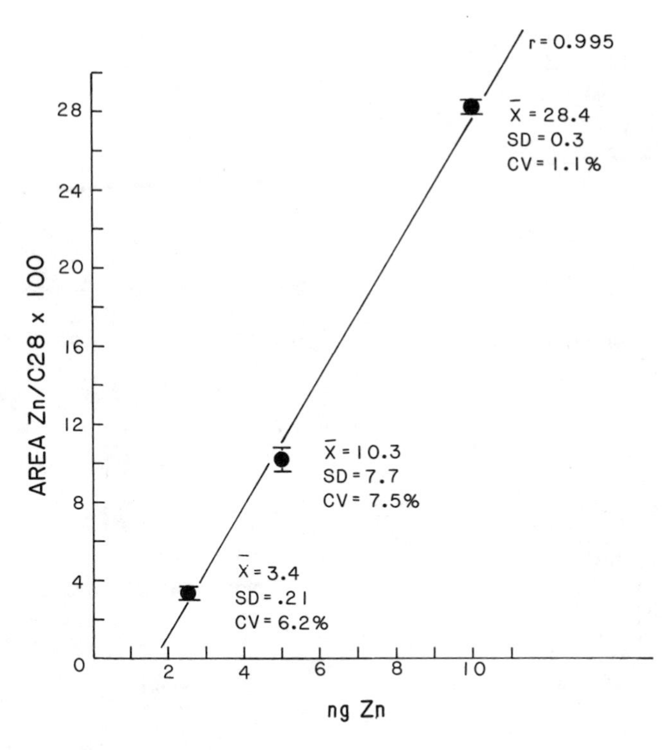

Figure 2. Standard curve obtained on gas chromatograph with hycrocarbon or internal standard.

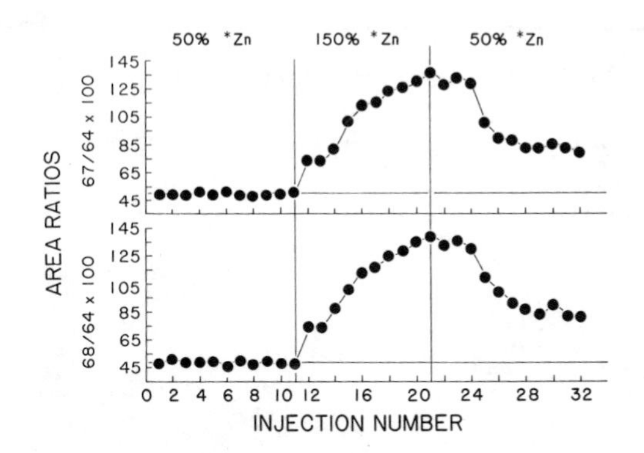

Figure 3. Zinc memory on gc/ms system. Theoretical figure would be a square wave.

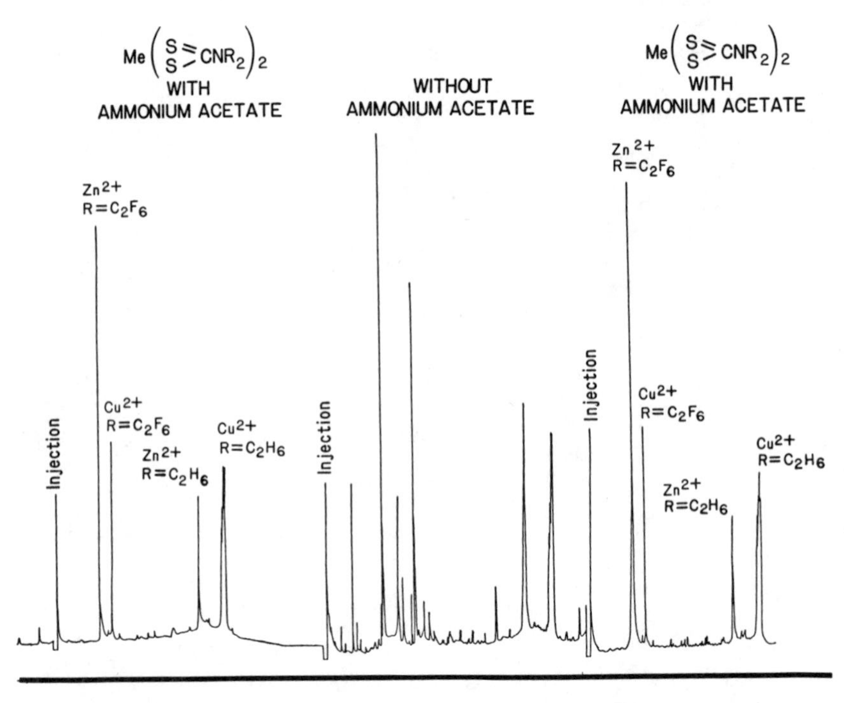

Figure 4. Three gas chromatograms of a mixture of zinc and copper trifluoroethyl and ethyl dithiocarbamate. Panels 1 and 3 are with ammonium acetate and panel 2 is without.

to the chelation solution. The second gc was of the same solution but where ammonium acetate was not added. The third gc was obtained from the original solution containing ammonium acetate. The results of this experiment are shown in Figure 4. The two chromatograms from the solution where the ammonium ion has been added show only 4 peaks representing the pair of bis-trifluoro chelates of copper and zinc and the corresponding pair of diethyl derivatives. The chromatogram from the sample without the ammonium ion shows multiple peaks. We interpret these results to indicate that the fifth ligand binding molecule has a stabilizing effect on the chelate complexes. The complexity and number of peaks seen in the center frame of Figure 4 present an additional problem in the interpretation of exactly what type of species is being chromatographed. With a simple exchange of ligands one would predict six or fewer peaks to be formed. The chromatogram shows at least eight major peaks and possibly more. This would lead one to consider that we are dealing with dimers of the metal chelates in the gas phase. These dimers would lead to at least ten peaks, and if the metals could also exchange this number would increase even more. Further study must be carried out on this aspect of the project.

More importantly, the fifth ligand apparently has greatly reduced the memory effect on the mass spectrometer. The results of a repeat of our earlier experiment where we inject natural zinc, enriched zinc and natural zinc are shown in Figure 5. The solid lines shown with each set represent the theoretical ratios. First, it is clear that we have essentially eliminated the memory and approached the theoretical ratios from either natural or enriched solution. This in itself represents a substantial step forward. Second, it is important to note that we have not completely eliminated the memory effect. A loading dose of 5 to 10 times the injected amounts is necessary to eliminate the final vestiges of earlier injections. Again, work is currently in progress to help understand the causes of this effect and to eliminate them.

Finally, in Figure 6 we show a calibration curve generated from multiple injections of known isotopic enrichments. In generating this curve we injected essentially a random selection of enrichments (i.e., high, natural, medium, etc.) so as to simulate a practical condition. The extrapolated line does have a slight curvature but we feel that it is good enough for us to begin our pilot patient studies.

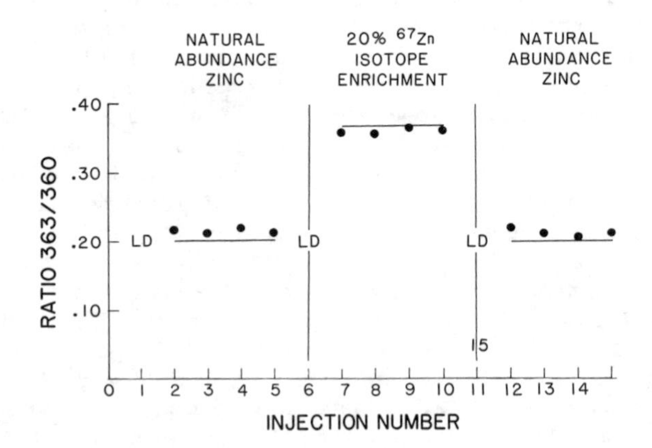

Figure 5. Repeat of experiment shown in Figure 3 with the addition of ammonium acetate to the solutions.

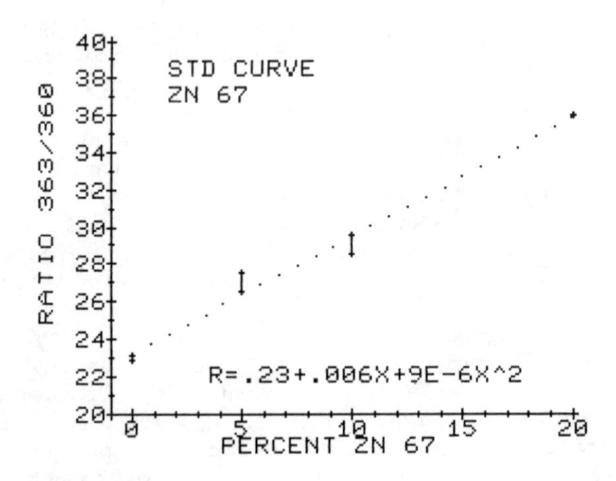

Figure 6. Measurement of Zinc 67 enrichment on gc/ms system using chelator and "fifth ligand" stabilizing compound.

Acknowledgments

I want to take this opportunity to thank the following people for their work in this project: Dr. Alain Favier, the National Institutes of Health (Grant Nos. RR01152, HD08315 and MCHS-Special Project 252) for partial financial support.

Literature Cited

1. Molokhia, M.; Sturniolo, G.; Shields, R.; Turnberg, L.A. Am. J. Clin. Nutr. 1980, 33, 881-6.
2. Aamodt, R.L.; Rumble, W.F.; Johnston, G.S.; Foster, D.; Henkin, R.I. Am. J. Clin. Nutr. 1979, 32, 559.
3. Foster, D.M.; Aamodt, R.L.; Henkin, R.I.; Berman, M. Am. J. Physiol. 1979, 237, R340-.
4. Spencer, H.; Vankinscott, V. Lewin, I.; Samachson, J. Man. J. Nutrition 1967, 86, 169-177.
5. Aamodt, R.L.; Rumble, W.F.; Johnson, G.S.; Markley, E.J.; Henkin, R.I. Fed. Proc. 1981, 403, 940.
6. Lonnerdal, B.; Ceberblad, A.; Sandstrom, B. Am. J. Clin. Nutrition 1983, 37, 695.
7. Janghorbani, M.; Young, V.R. Am. J. Clin. Nutrition 1980, 33, 2021-30.
8. Janghorbani, M., Young, V.R., in "Advances in Nutritional Research"; Draper, H. H., Ed.; Plenum: New York, 1980; Vol. III, pp. 127-155.
9. King, J.C.; Raynolds,W.L.; Margen, S. Am. J. Clin. Nutrition 1978, 3, 1198-1203.
10. Johnson, P.E. XII International Congress of Nutrition, (1981).
11. Johnson, P.E. J. of Nutrition 1982, 112, 1414.
12. Turnlund, J.R.; King, J.C., in "Nutritional Bioavailability of Zinc"; Inglett, G., Ed.; ACS SYMPOSIUM SERIES No.120, American Chemical Society: Washington, D.C., 1983; p. 31.
13. Schwartz, R.; Spencer, H.; Wentworth, R.A. Clinica Chimica Acta. 1978, 87, 265-73.
14. Hui, K.S.; Davis B.A.; Boulton, A.A. Neurochem. Res. 1977, 2, 495-506.
15. Hachey, D.L.; Blais, J.C.; Klein, P.D. Anal. Chem. 1980, 52, 1131-5.
16. Risby, T.H. Environmental Health Perspectives 1980, 36, 39-46.
17. Davis, B.A.; Hui, K.S.; Boulton, A.A. Adv. Mass Spectrom. Biochem. Med. 1977, 2, 405-18.
18. Prescott, S.R.; Campana, J.E.; Risby, T.H. Anal. Chem 1977, 49, 1501-4.
19. Kraus, K.A.; Moore, G.E. J. Am. Chem. Soc. 1958, 75, 1460-2.
20. Radecki, A.; Halkiewicz, J. Journal of Chrom. 1980, 187, 363.

21. Burgett, C.A. The Gas Chromatography of B-Diketonates Separations and Purification Methods 1976, 5, 1-32.
22. Uden, P.C.; Blessel, K. Inorganic Chemistry, 1973, 12, 352-356.
23. Uden, P.C.; Henderson, D.E. The Analyst, 1977, 102, 889-916.
24. Radecki, A.; Halkiewicz, J.; Grzybowski, J.; Lamparczyk, H. J. of Chrom. 1978, 151, 259-62.
25. Tavlaridid, A.; Neeb, R. Z Anal. Chem., 1976, 282, 17-19.
26. Coates, E.; Rigg, B.; Saville, B.; Skelton, D. J. Chem. Society 1965, 5613.

RECEIVED December 23, 1983

10

Stable Isotopes of Iron, Zinc, and Copper

Used to Study Mineral Absorption in Humans

PHYLLIS E. JOHNSON

U.S. Department of Agriculture, Agricultural Research Service,
Human Nutrition Research Center, Grand Forks, ND 58202

Stable isotopes of minerals are useful tools for studying metal absorption and bioavailability in humans. Analyses of stable isotope enrichment in biological samples are much more complex than radioisotope measurements. Conventional electron-impact mass spectrometry (EI-MS) of metal chelates provides a relatively rapid method for measurement of enrichment levels. Precision of EI-MS isotope ratio measurements in metal-tetraphenylporphyrins is about 2%. This is usually adequate for metal apsorption studies using the fecal monitoring method. Uncertainty in fractional absorption values calculated from EI-MS data is 0.005-0.03 absorption units. This method was used to measure Fe, Zn, and Cu absorption in adult men participating in metabolic studies, and in infants. Addition of picolinic acid to a tryptophan-limited diet increased absorption of zinc but did not affect copper absorption. Breast-fed infants appeared to absorb more Zn, Fe, and Cu than bottle-fed infants. Post-absorptive excretion of all three metals occurred from 10 days to six weeks after isotope feeding. Higher levels of metal absorption were associated with less post-absorptive excretion.

Determination of human requirements for dietary trace minerals necessarily includes knowledge of the factors which affect the availability of minerals for absorption. Interactions of dietary minerals with organic constituents of the diet and with other minerals are complex. Careful study of mineral absorption in human subjects is required to delineate dietary requirements and the factors which affect them.

In the last five years stable isotopes of many metals have proven to be useful in studying mineral absorption and metabolism in human subjects. There is growing reluctance to use radioisotopes for such experiments, and they are contraindicated in some populations. Stable isotopes are beginning to fill this gap.

At present there is no standard method for measurement of stable isotope enrichment in biological samples. The diversity of analytical approaches reflects the varied backgrounds of the investigators involved, and the instrumentation which has been available to them. Nutrition studies tend to generate many samples, and a dedicated mass spectrometer or other facility is advantageous.

We wished to develop a method for measuring isotopic enrichment of iron, zinc, and copper in biological samples using a conventional electron impact ionization mass spectrometer. In order to achieve enough volatility to vaporize a sample into the mass spectrometer, it is necessary to prepare metal chelates. Various ligands have been used for this purpose, including 2,4-pentanedione (acetylacetone) (1), fluorinated diones (2), diethyldithiocarbamates (2), and *meso*-tetraphenylporphyrin (TPP) (3). The method used in this laboratory utilizes metal-TPP chelates.

Chelation of copper with tetraphenylporphyrin was first reported as a technique for measuring copper in rat brain by isotope dilution (3). This method appeared to be adaptable for measurement of isotopic enrichment in other biological samples. Tetraphenylporphyrin reacts quickly and quantitatively to form chelates with nearly all of the transition metals (4). The mass spectrum of a TPP chelate is dominated by the molecular ion cluster (Figure 1); the molecular ion is also the base peak, so it is well-suited for quantitation of isotope ratios.

Materials and Methods

Preparation of Chelates. The procedure we use is adapted from that used by Hui et al. (3) to measure copper in rat brain. Samples of feces, approximately 0.25 g, were ashed in a low temperature asher, dissolved in concentrated HCl, and applied to Bio-Rad AG-1 anion exchange columns. [Mention of a trademark or proprietary produce does not constitute a guarantee or warranty of the product by the U.S. Department of Agriculture, and does not imply its approval to the exclusion of other products that may also be suitable.] The various metals were then separated by elution with successively lower concentrations of HCl. Recovery of metals from the columns was 97.6±2.7% for Fe, 100±4.2% for Cu, and 93.7±1.8% for Zn (5). Each metal fraction was then refluxed separately with TPP in dimethylformamide to form the metal chelate (5).

Hui et al. did not separate metals in their samples before

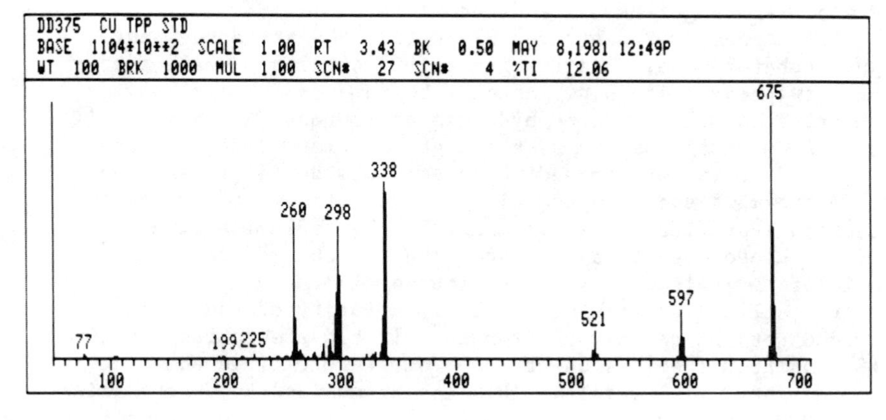

Figure 1. Mass spectrum of copper tetraphenylporphyrin.

chelation; instead they performed a thin-layer chromatography (TLC) step after formation of the metal-TPP complexes. We found that this procedure worked well for synthetic mixtures of metals, but not for fecal samples. We have better results when the metals are first separated and then refluxed separately with the TPP. The Cu-TPP samples were further purified by TLC after chelation to remove residual Zn-TPP, but TLC of the Zn- or Fe-TPPs did not seem to improve the quality of the mass spectrum. Although Hui et al. reported separation of metal-TPP mixtures in eleven solvent systems and on four different adsorbents (6), we achieved good separation with only one system, toluene:ligroin:acetic acid:water (35:65:15:85) on Pierce silica gel LK5D precoated plates.

It is necessary to carefully purify the tetraphenylporphyrin before use. Reagent grade TPP contains a small amount of tetraphenylchlorin, which must be removed. Tetraphenylchlorin contains two more hydrogen atoms than TPP, and it will also form metal complexes, although to a lesser degree than TPP. If it is not removed it causes substantial interference with the mass spectrum and with the measurement of isotopic ratios. Purification is accomplished by refluxing with 2,3-dichloro-5,6-dicyanobenzoquinone in ethanol-free chloroform, followed by filtration on alumina and recrystallization from methanol (7). Purity of the final product can be determined from the visible absorption spectrum (8) and from the mass spectrum. Commercial TPP is sometimes contaminated with metals. This can be avoided by synthesizing the TPP from pyrrole and benzaldehyde in refluxing propionic acid (9)

Mass Spectrometry. The mass spectrometer used was a DuPont DP-102. Spectra were recorded in the electon-impact ionization mode, at a source temperature of 220° and an ionization potential of 70eV, with a probe temperature of 300°.

Samples spiked with known amounts of stable isotopes ^{54}Fe, ^{57}Fe, ^{65}Cu, ^{67}Zn, ^{70}Zn were prepared, and the measured ion intensities in the mass spectra were compared to theoretical values. The response of the mass spectrometer was linear over a wide range of enrichment levels (5). A computer program included in the mass spectrometer software package was used to calculate the theoretical values for peak intensities in the mass spectra. Corrections for isotopic composition of the ligand (^{13}C, ^{15}N, ^{2}H) were made as described previously (5).

Calculation of Absorption. Absorption of isotope doses by human subjects was calculated by fecal monitoring (5), that is, by determining the amount of stable isotope ingested compared to the amount excreted in the feces in excess of the natural abundance. The fractional absorption $A = (I-F \cdot E)/I$ where I is the amount of isotope ingested, F is the total metal in the feces, and E is the enrichment of the isotope above natural abundance. If multiple fecal collection periods are used, then $A = (I - \Sigma F_i E_i)/I$.

Precision. The precision of the absorption value depends upon the precision of I, F, and E measurements. This method for isotopic enrichment measurements by mass spectrometry has a precision of 2%, as does the measurement of F by atomic absorption. This precision is adequate for absorption and bioavailability studies with zinc and copper (Table I), since zinc and copper absorption are in the range of 30-70%. Only fairly large changes in iron absorption can be discerned, because non-heme iron absorption is typically less than 10%. This may not be a serious problem in bioavailability studies, since it is doubtful that very small changes in iron absorption from single foods are biologically significant.

Table I. Typical Uncertainties in Absorption Values with Measurement Precision of 2%

Fractional Absorption	Uncertainty in A (σA)
.05	.027
.10	.025
.55	.013
.77	.006
.90	.003

Absorption of stable zinc measured by this method compared to absorption of ^{65}Zn measured by whole body counting are in agreement within the limits of error of the two methods (10).

Human Studies. Male volunteers who were participating in studies at the metabolic unit of the Grand Forks Human Nutrition Research Center (GFHNRC) gave consent after being informed of the purpose of the research and its potential hazards. Consent was given by parents of infants participating in out-patient studies. This project was approved by the Human Studies Committees of the University of North Dakota School of Medicine and of the USDA-Agricultural Research Service. Informed consent and experimental procedures were consistent with the Declaration of Helsinki. All volunteers were in good health except an elderly subject who was studied because he had gluten enteropathy. All volunteers living in the metabolic unit were fed constant mixed diets prepared from conventional foods. All subjects were chaperoned when they left the metabolic unit to prevent ingestion of unauthorized foods or loss of excreta samples.

Stable isotopes of iron, zinc, and copper have been given to subjects in a variety of studies at our Center. Standard doses were 4 mg ^{54}Fe, 4 mg ^{67}Zn and 2 mg of ^{65}Cu for adult subjects. Isotopes were fed in a single dose in juice at breakfast or in Trutol (a flavored glucose solution). While doses of this size are not truly tracer doses, they are in the physiological range of intake. Recommended daily intakes are 10 mg Fe, 15 mg Zn and 2-3 mg Cu for adult males (11). Table II shows the average absorption values obtained from subjects who were consuming diets adequate in all minerals and other dietary components.

Table II. Typical Values for Absorption of Stable Fe, Zn, Cu by Adult Males

	% Absorption	
	with food	without food
Fe	14 ± 18 (7)*	5 ± 3 (6)
Zn	44 ± 15 (12)	72 ± 24 (8)
Cu	59 ± 13 (7)	78 ± 12 (9)

*mean ± standard deviation (number of subjects)

The values for iron absorption with food were for iron given in a breakfast meal containing orange juice, which explains why they were somewhat higher than the values without food, since ascorbic acid is known to enhance non-heme iron absorption (12). Absorption of non-heme iron absorption from food without added ascorbate is usually 10% or less (13). The values for zinc absorption in the absence of food were in the same range as those determined by Aamodt et al. (14) using ^{65}Zn. Absorption of ^{65}Zn given with food tends to be lower, as these values are (15,16). Thus results obtained using this method appear comparable to absorption values measured by other techniques.

Some Typical Uses of Stable Isotope Absorption Studies

Tryptophan and Zinc Absorption. Doses of 4 mg ^{67}Zn and 2 mg ^{65}Cu were given in Trutol to subjects consuming a diet in which tryptophan was the limiting amino acid. Absorption of Zn and Cu was also measured when they received the same diet plus a picolinic acid supplement. Picolinic acid is a metabolite of tryptophan which is thought to enhance zinc absorption (17). Picolonic acid was not present in the isotope dose. The diets contained 0.8-1.2 mg Cu/day and 2.7-3.6 mg Zn/day (by analysis). Total tryptophan content of the diet was approximately 250 mg/day (calculated). Picolinic acid supplementation was at a level of 10 mg/day.

The picolinic acid supplement increased zinc absorption in all three subjects, but had little or no effect on copper absorption (Table III).

Table III Stable Zinc and Copper Absorption by Men Consuming a Tryptophan-Limited Diet With and Without Picolinic Acid Supplementation

	% Zn Absorption		% Cu Absorption	
Subject No.	Lo Trp	Lo Trp + PA	Lo Trp	Lo Trp + PA
2055	33	55	59	58
2056	62	82	76	--
2070	47	52	53	57

Gluten-sensitive enteropahy. A subject with gluten enteropathy was tested. While hospitalized with this disease, he had been found to have a plasma Zn of < 5 μg/dl (18). It was desired to determine if he had abnormal Zn absorption or was Zn-deficient due to the diarrhea and malnutrition of his disease. We found that he had normal absorption of Zn, Fe, and Cu from Trutol (Table IV) while on a gluten-free diet.

Table IV. Metal Absorption by a Subject with Gluten-Sensitive Enteropathy

	% Absorption		
	Zn	Fe	Cu
Subject	75	4.5	82
Controls	72 ± 26*	5.1 ± 3.7	77 ± 13

*Mean ± Standard Deviation

Normal Zn absorption was confirmed by whole-body counting of ^{65}Zn. However, the subjects retention of ^{65}Zn was extremely short, and turnover of ^{65}Zn was faster when he was free-living than when he resided on the metabolic unit and consumed a carefully-controlled gluten-free diet (19).

Infants. One advantage of stable isotopes is that they can be given to infants, children, and pregnant women. We have done some pilot studies with breast-fed and bottle-fed babies to whom we have given stable Fe, Zn, and Cu. Stable isotopes of all three metals (2 mg ^{54}Fe, 2 mg ^{67}Zn, 1 mg ^{65}Cu) were mixed with breast milk or formula (SMA) and fed to 2-month old infants. All stools were collected for four days. All three metals appear to be better absorbed by the breast-fed infants (Table V). This is consistent with reports for ^{65}Zn absorption by adults of 41±9% from human milk and 31±7% from a humanized cow's milk formula (mean ± SD) (20). Iron absorption from breast milk using ^{59}Fe has been reported to be 10-15% in adults (21,22) and 49% in 6 month old infants (23).

Table V. Absorption of Zn, Fe, Cu From Breast Milk or Formula

	% Absorption		
	Zn	Fe	Cu
Breast-fed	50 ± 4 (5)*	30 ± 11 (2)	50 ± 3 (5)
Bottle-fed	27 ± 7 (3)	20 ± 5 (2)	30 ± 10(2)

*mean ± standard error of the mean (number of subjects)

This pilot study with a small number of infants does not permit firm conclusions about metal absorption in infants. However, we have learned that bottle-fed infants tolerate the metal isotope solutions better than breast-fed infants. In a subsequent study, many (6/20) breast-fed babies spit up the milk-isotope mixture within 30 minutes of feeding. Two others refused it entirely. In contrast, all bottle-fed babies (12/12) drank the isotope-formula mixture and none had emesis. The doses of isotopes could probably be reduced by at least half without sacrificing analytical sensitivity, and this would probably reduce gastrointestinal intolerance.

Fecal Excretion of Isotopes. One of our more recent activities has been to examine the patterns of excretion of stable Zn, Cu, and Fe in the feces. Janghorbani reported that it is necessary to collect only five stools in order to recover all of the unabsorbed isotope (24-26). This seemed contrary to our experience, so we undertook a systematic examination of our stool data.

In many cases more than one pulse of isotope was excreted in the feces after a single dose of stable isotope (Figure 2). We observed this with more than 20 subjects. The phenomonen is reproducible for separate doses of isotope in the same subject, and for three different metals, analyzed independently. To my knowledge this has not been reported previously, although Björn-Rasmussen et al. (27) and Lykken (28) have observed post-absorptive excretion of iron using whole-body counting of ^{59}Fe. They report excretion of "absorbed" iron more than 14 days after ingestion of the ^{59}Fe. The amount of post-absorptive excretion (PAE) they observed seemed to be related to serum

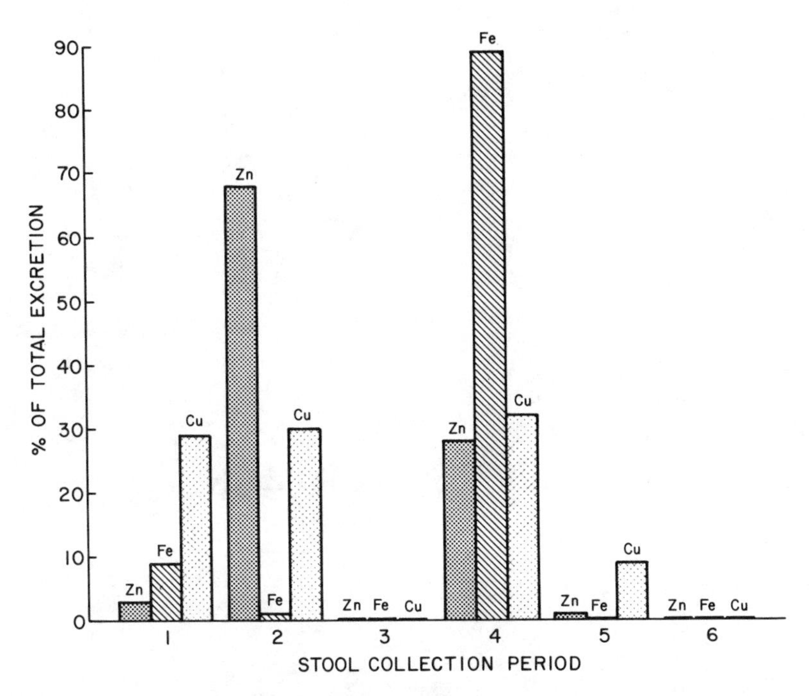

Figure 2a. Excretion of stable iron, zinc, and copper by two human subjects.

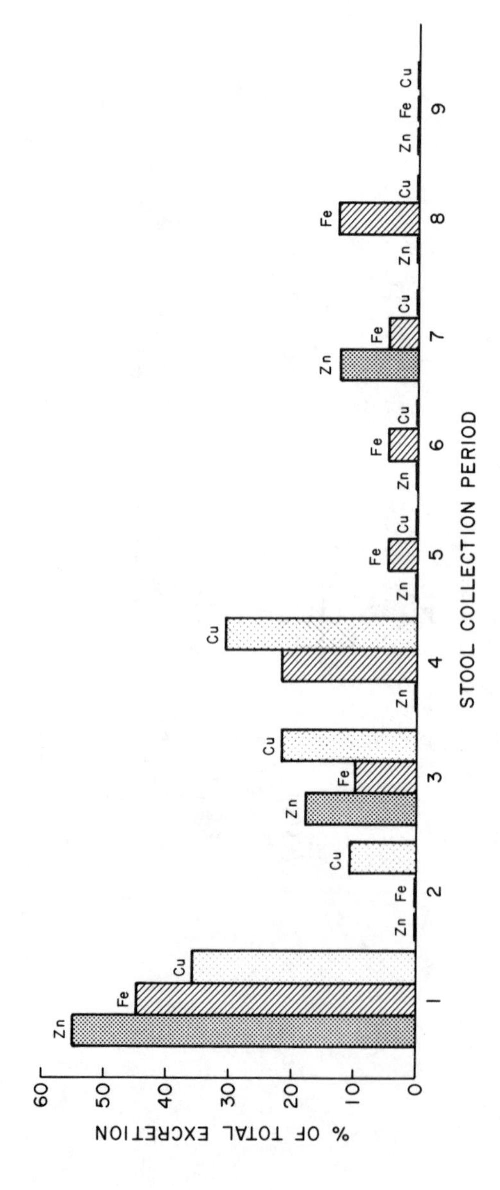

Figure 2b. Excretion of stable iron, zinc, and copper by one human subject at one time.

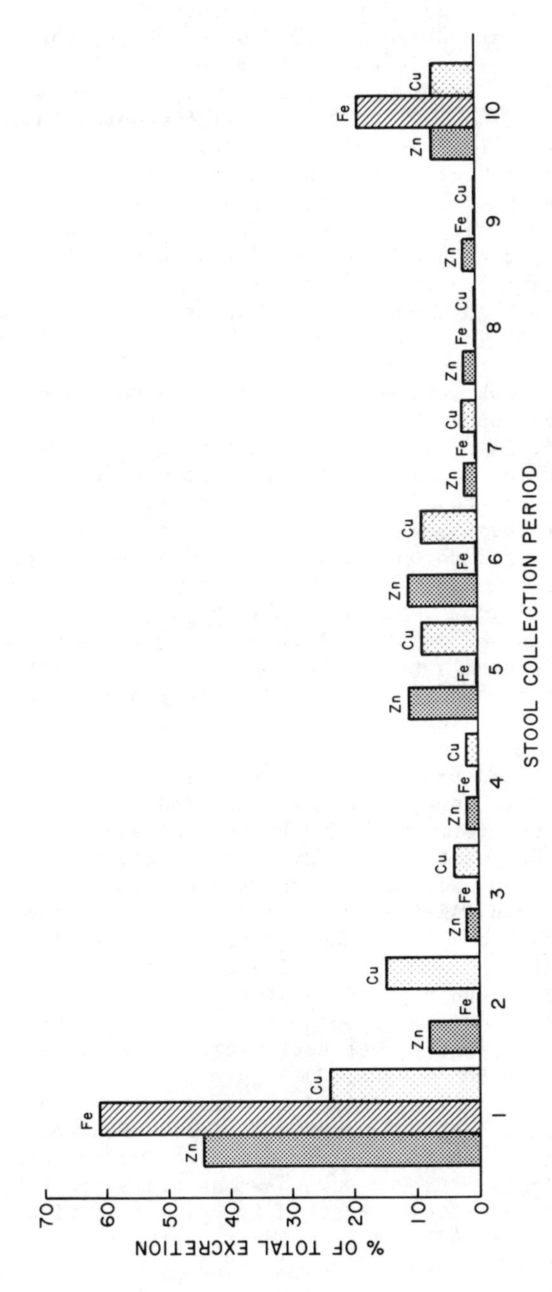

Figure 2c. Excretion of stable iron, zinc, and copper by the same human subject as in Figure 2b at a different time.

ferritin, or in other words, to body iron stores. We postulate that the first pulse of isotope excretion represents unabsorbed isotope and that subsequent pulses represent "post-absorptive" excretion. If this is true, it obviously has implications for the size of the fecal pool used in the fecal monitoring method. A very large (e.g., 21-day) pool might underestimate absorption considerably. Daily sampling or analysis of individual stools has the drawback of increasing the sample load considerably. The work reported here used three-day collections as a compromise between these two extremes.

The number of stools needed for a complete collection of unabsorbed isotope (the first pulse of isotope excreted) was clearly more than five for a large part of our population (Figures 3-5). The figures show the number of stools for complete collection of the first pulse of isotope excreted. All of these subjects were consuming controlled diets made from conventional foods, while Janghorbani's subjects were consuming formula (24,25) or semi-synthetic (26) diets. The number of days needed for complete collection of unabsorbed isotope ranged from one to 33 days. A 21-day stool collection was enough to recover unabsorbed material in 90% of our subjects. However, it is important to note that the feces from the 21-day period cannot be analyzed in a single pool. Feces must be analyzed individually or in small (e.g., 3 day) pools, in order to observe the end of the first pulse of isotope excretion, which is presumably the unabsorbed isotope. A single too-short stool collection period may overestimate absorption. If post-absorptive excretion occurs, a single, extensively long stool collection period will underestimate absorption.

As mentioned previously, Björn-Rasmussen and Lykken both reported a relationship between post-absorptive excretion of iron and serum ferritin. The higher the subject's serum ferritin, the greater the body iron stores, and the more post-absorptive excretion which was observed. For uniformity, we have defined PAE-9 as the percentage of the dose of isotope which was excreted more than 9 days after the dose. Nine days comprise three 3-day stool collection periods. Bjorn-Rasmussen used a 10-day cutoff point (29).

In this group of subjects there was no relationship between ferritin and Fe-PAE-9, but serum iron was correlated with PAE-9 with an r of 0.508 at $p<0.02$ (Figure 6). That is, subjects with higher serum iron tended to excrete more iron 9 days or more after the dose. Possibly, subjects with better iron status have some hold-up of iron in the gut and excrete it later. Likewise, Fe-PAE-9 was exponentially correlated with the percentage of iron absorbed from the dose with $r^2=0.35$ and a $p<0.005$ (Figure 7). This means that the more iron which was absorbed, the less which was excreted at day 10 or later.

A similar relationship between percentage of Zn absorption and Zn-PAE-9 (Figure 8) was observed except that the

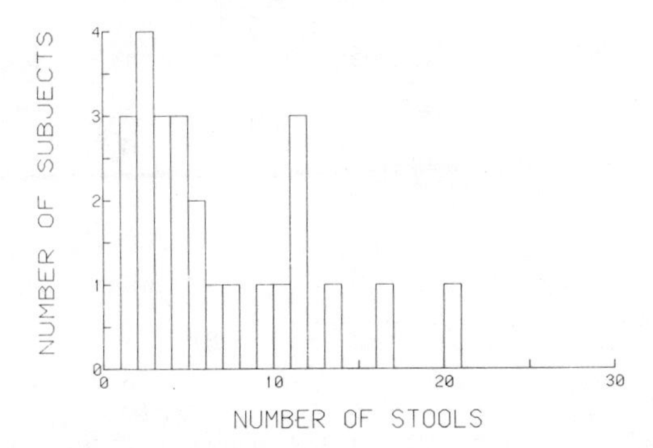

Figure 3. Number of stools required to collect unabsorbed stable zinc.

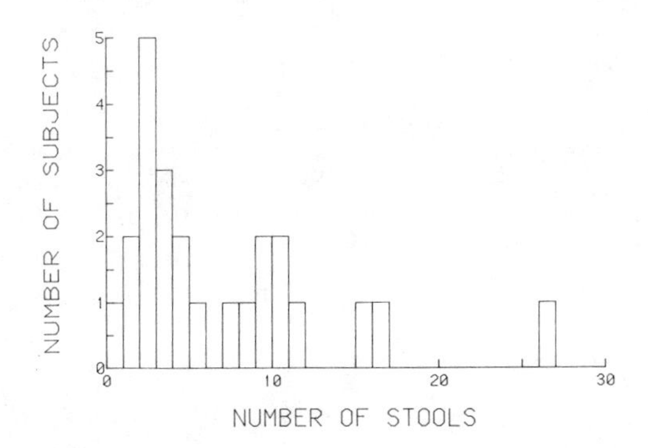

Figure 4. Number of stools required to collect unabsorbed stable copper.

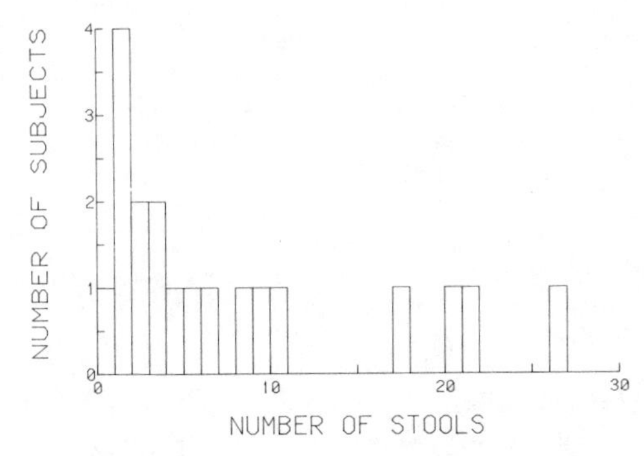

Figure 5. Number of stools required to collect unabsorbed stable iron.

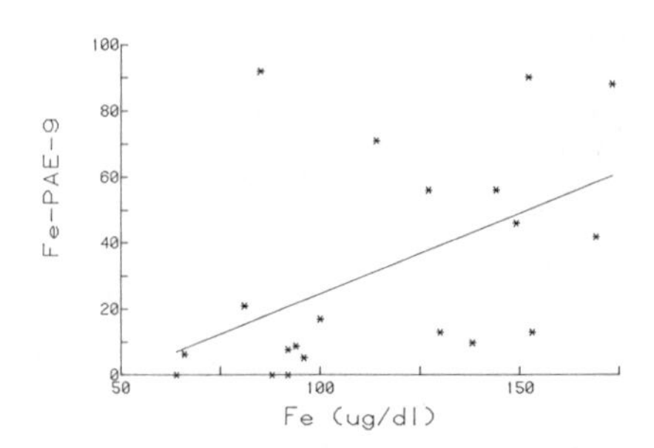

Figure 6. Relationship between serum iron and post-absorptive excretion (PAE-9) of iron.

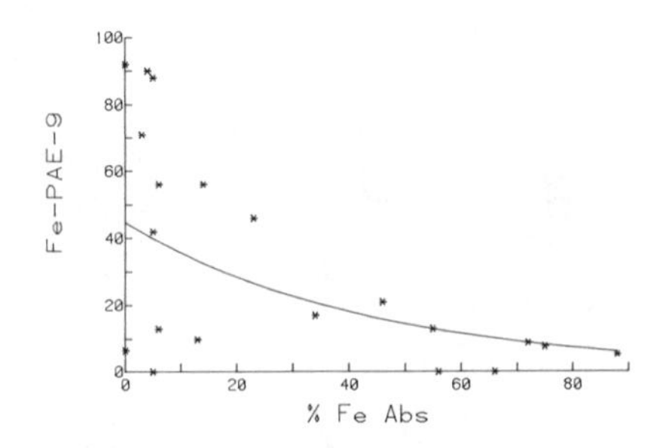

Figure 7. Relationship between percent iron absorption and post-absorptive excretion (PAE-9) of iron.

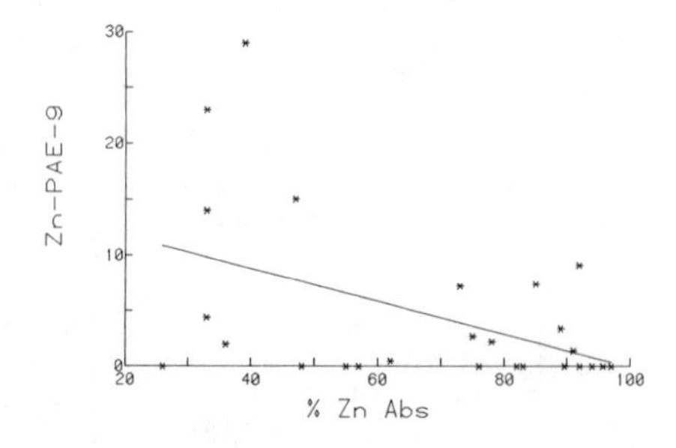

Figure 8. Relationship between percent zinc absorption and post-absorptive excretion (PAE-9) of zinc.

relationship was linear rather than exponential. Here $r = -0.46$ with a $p<.02$. Serum Zn was not related to Zn-PAE-9.

For Cu there was no relationship between serum Cu or ceruloplasmin and Cu-PAE-9. However, the percentage of the total excreted isotope which was excreted after nine days, which we call E-9, was linearly related to the % Cu absorption with $r = 0.43$ at $p<0.03$.

Post-absorptive excretion of zinc and copper has not previously been reported. Presumably the same or similar mechanisms are involved in the excretion of iron, zinc, and copper, since the excretion of the three metals appears to be synchronous in most subjects. Post-absorptive excretion of iron goes on for four to five weeks after an oral dose of ^{59}Fe (29). Using stable isotopes of Fe, Zn, and Cu, we have observed post-absorptive excretion of all three metals for up to six weeks after the isotope dose.

Several hypotheses have been proposed to explain post-absorptive excretion of iron (29), but few data are available to differentiate among them. It seems unlikely that post-absorptive excretion is due only to sloughing of isotopically labelled enterocytes, since their lifespan of a few days would not account for excretion of isotope for several weeks following ingestion. Possibly, Fe, Zn, and Cu are excreted bound to metal-carrying proteins which can be reabsorbed lower in the gut in an enteroenteric recirculation process (29). Alternatively, there may be storage of metal in gut macrophages (29-31) perhaps followed by excretion through goblet cells (29,31). It is also possible that post-absorptive excretion involves excretion of metal which ws completely absorbed and subsequently passed back into the mucosa from blood. Several mechanisms are known for transfer of iron from blood to the mucosal cells (32).

Post-absorptive excretion of zinc may involve zinc which has entered the enteropancreatic circulation, since not all of the zinc in pancreatic secretions is reabsorbed (33).

It is possible that excretion of isotopes in more than one pulse is due to peristaltic reflux or variations in fecal flow for different components of feces. Turnlund (34) has observed excretion of polyethyleneglycol (PEG), a non-absorbable marker, in a multiphasic pattern. However, it seems unlikely that this would explain excretion of isotope as long as six weeks after ingestion.

All of our absorption calculations, such as those presented above, have been based on the assumption that the first pulse of excretion represented unabsorbed isotope and subsequent excretion was "post-absorptive." These absorption values agree with metal absorption determined by other means, such as whole-body counting.

Arriving at an explanation for multiphasic isotope excretion patterns will doubtless require much further work.

In any event, it is clear that for the subjects consuming conventional diets, collection of five stools was insufficient for recovery of all unabsorbed isotope.

Some of these questions probably cannot be resolved with stable isotope studies. However, stable isotopes may well be of use in identifying conditions which affect post-absorptive metal excretion.

Summary

Stable isotopes of minerals are safe and convenient to use for bioavailability studies in humans. Chelation of metals to form volatile complexes makes it possible to use ordinary EI/MS instrumentation for measurement of isotopic enrichment. Stable isotopes can be complementary to radioisotopes in some of experiments. Observations of fecal isotopic excretion are inherent to bioavailability and absorption studies using stable isotopes and are unlikely to be made in studies employing low level radioisotopes.

Acknowledgments

The author thanks Cheryl Stjern for technical assistance, David B. Milne, Ph.D.; Janet R. Mahalko, M.S., R.D.; and LuAnn K. Johnson, M.S.

Literature Cited

1. Miller, D.D.; VanCampen, D. J. Nutr. 1979, 32, 2354-2361.
2. Hachey, D.L.; Blais, J.-C.; Klein, P.D. Anal. Chem. 1980, 52, 1131-1135.
3. Hui, K.-S.; Davis, B.A.; Boulton, A.A. Neurochem. Res. 1977, 2, 495-506.
4. Davis, B.A.; Hui, K.-S.; Boulton, A.A. Adv. Mass Spectrom. Biochem. Med. 1976, 2, 405-418.
5. Johnson, P.E. J. Nutr. 1982, 112, 1414-1424.
6. Hui, K.-S.; Davis, B.A.; Boulton, A.A. J. Chrom. 1975, 115, 581-586.
7. Barnett, G.H.; Hudson, M.F.; Smith, K.M. J. Chem. Soc. Perkin Trans. I 1975, 1401-1403.
8. Rouseau, K.; Dolphin, D. Tetr. Lett. 1974, 48, 4251-4254.
9. Adler, A.P.; Longo, F.R.; Finarelli, J.D.; Goldmaiher, J.; Assour, J.; Korsakoff, L. J. Org. Chem. 1967, 32, 476.
10. Johnson, P.E.; Lykken, G.; Mahalko, J.; Milne, D.; Inman, L.; Sandstead, H.H.; Garcia, W.J.; Inglett, G.E. in "The Maillard Reaction in Foods and Nutrition"; Waller, G.R.; Feather, M.S., Eds.; ACS SYMPOSIUM SERIES No. 215, American Chemical Society: Washington, D.C., 1983; pp. 349-360.

11. National Academy of Sciences. National Research Council. 1980, Recommended Dietary Allowances.
12. Monsen, E.R.; Hallberg, L.; Layrisse, M.; Hegsted, D.M.; Cook, J.D.; Mertz, W.; Finch, C.A. Am. J. Clin. Nutr. 1978, 31, 134-141.
13. Cook, J.D. Fed. Proc. 1977, 36, 2028-2032.
14. Aamodt, R.L.; Rumble, W.F.; Johnston, G.S.; Markley, E.J.; Henkin, R.I. Am. J. Clin. Nutr. 1981, 34, 2648-2652.
15. Sandström, B.; Arvidsson, B.; Cederblad, Å.; Björn-Rasmussen, E. Am. J. Clin. Nutr. 1980, 33, 739-745.
16. Sandström, B.; Cederblad, Å. Am. J. Clin. Nutr. 1980, 33, 1778-1783.
17. Evans, G.W.; Johnson, E.C. J. Nutr. 1980, 110, 1076-1080.
18. Milne, D., Personal Communication.
19. Lykken, G.I., Personal Communication.
20. Sandström, B.; Cederblad, Å.; Lönnerdal, B. Am. J. Dis. Child. 1983, 137, 726-729.
21. McMillan, J.A.; Landaw, S.A.; Oski, F.A. Pediatrics 1976, 58, 686-691.
22. McMillan, J.A.; Oski, F.A.; Lourie, G.; Tomarelli, R.M.; Landaw, S.A. Pediatrics 1977, 60, 896-900.
23. Saarinen, U.M.; Siimes, M.A.; Dallman, P.R. J. Pediatrics 1977, 91, 36-39.
24. Janghorbani, M.; Ting, B.T.G.; Young, V.R. J. Nutr. 1980, 110, 2190-2197.
25. Solomons, N.W.; Janghorbani, M.; Ting, B.T.G.; Steinke, F.H.; Christensen, M.; Bijlani, R.; Istfan, N.; Young, V.R. J. Nutr. 1982, 112, 1809-1821.
26. Christensen, M.J.; Janghorbani, M.; Steinke, F.H.; Istfan, N.; Young, V.R. Br. J. Nutr. 1983, 50, 43-50.
27. Björn-Rasmussen, E.; Carneskog, J.; Cederblad, Å. Scand. J. Haematol. 1980, 25, 124-126.
28. Lykken, G.I., Personal Communication.
29. Björn-Rasmussen, E. Lancet 1983, i, 914-916.
30. Cattan, D. Lancet 1983, ii, 106.
31. Schreiner, B.; Refsum, S.B. Lancet 1983, i, 1385.
32. Linder, M.; Munro, H.M. Fed. Proc. 1977, 36, 2017-2023.
33. Mataseshe, J.W.; Phillips,, F.S.; Malgelada, J.-R.; McCall, J.T. Am. J. Clin. Nutr. 1980, 33, 1946-1953.
34. Turnlund, J. Personal Communication.

RECEIVED January 31, 1984

11

^{13}C-Enriched Substrates for In Situ and In Vivo Metabolic Profiling Studies by ^{13}C NMR

J. R. BRAINARD, J. Y. HUTSON, and R. E. LONDON

Isotope and Nuclear Chemistry Division, University of California, Los Alamos National Laboratory, P.O. Box 1663, MS J515, Los Alamos, NM 87545

N. A. MATWIYOFF

Isotope and Nuclear Chemistry Division, University of California, Los Alamos National Laboratory, P.O. Box 1663, MS J515, Los Alamos, NM 87545 and University of New Mexico/Los Alamos NMR Center for Non-Invasive Diagnosis, Albuquerque, NM 87131

The application of stable isotopes and NMR to the study of metabolism and its regulation in living systems has unique advantages over traditional biochemical or physiological techniques. From the concentration and labeling patterns observed in labeled products and intermediates, information concerning the regulation and relative contribution of metabolic pathways can be obtained. In experiments investigating the regulation of gluconeogenesis in perfused hamster livers, the fluxes of carbon-13-labeled alanine through multiple metabolic pathways leading to glucose and glycogen have been determined and modulation of these metabolic pathways by reducing substrates demonstrated. In addition, the effects of reducing substrates on the intracellular redox potential of the perfused liver have been observed by ^{13}C NMR.

For over a decade, the combination of ^{13}C-labeled substrates and ^{13}C nuclear magnetic resonance (NMR) spectroscopy has been used to study product precursor relationships and the activity of metabolic pathways in microorganisms (1-3). In these studies, the ^{13}C-labeled substrate is added to the nutrient medium of the microorganism and at some later time, metabolic products suspected of containing the label (often on the basis of prior studies with the radioactive isotope, ^{14}C) are isolated from the medium or extracted from the microorganism, and the distribution of the ^{13}C label at specific sites is quantified by high resolution ^{13}C NMR or ^{1}H NMR spectroscopy of an extract containing the metabolites of interest. Recently ^{13}C NMR spectroscopy in conjunction with ^{13}C labeling has become an important tool for studying metabolism in intact, living tissue (4-6).

0097-6156/84/0258-0157$06.00/0

The hallmark of the method is that metabolites can be measured repetitively and non-destructively in intervals approaching real time, without extraction of the metabolites from the tissue. The development of large bore magnets and surface coils (7) is now allowing the non-invasive study of metabolism of ^{13}C-labeled compounds in live animals (8). We anticipate that soon ^{13}C NMR spectroscopy and ^{13}C-labeled substrates will be used for non-invasive studies of metabolism in humans, extending the range of physiological and biochemical disorders that already are being actively addressed in ^{31}P NMR studies of humans (5,9). The ^{13}C methodology will be particularly important in the study of nutritional disorders and nutritional requirements under stress since many metabolites of interest do not contain phosphorus atoms and cannot be studied directly by ^{31}P NMR.

Survey of Recent ^{13}C NMR Metabolic Studies

The status of *in situ* and *in vivo* ^{13}C NMR studies of ^{13}C-labeled substrates has been reviewed within the last year (4,5,9). We survey briefly some of the notable recent developments since those reviews appeared. Sillerud and Shulman (10) were able to quantitate the *natural* abundance ^{13}C signals of glycogen *in situ* within the perfused livers of rats and were able to compare the relaxation properties (T_1, T_2, and Nuclear Overhauser effect) of glycogen *in situ* and *in vitro*. In spite of the high molecular weight of glycogen, high resolution ^{13}C NMR spectra were obtained and the relaxation parameters were found to be identical *in situ* and *in vitro*, suggesting a high degree of internal motion of segments of the polymer and a pseudo-isotropic overall motion for the glycogen itself. These studies have been extended to the *in vivo* investigation of hepatic glycogen synthesis in rabbits (11) and *in vivo* guinea pig heart metabolism (12). In the study of guinea pig heart metabolism, the time course of glycogen synthesis after intravenous injection of [1-^{13}C]-D-glucose was followed serially and the effects of anoxia on glycogen mobilization were monitored. In like manner, Behar *et al.* (13) have accomplished *in vivo* serial ^{13}C NMR studies of hypoxia in rabbit brain after intravenous infusion of [1-^{13}C]-D-glucose by monitoring the rise and fall of the flow of label into the C-3 carbon of lactate with induction and termination of hypoxia. Of special interest is the recent report (14) of the acquisition of localized ^{13}C NMR spectra (at natural ^{13}C abundance) of the head and body of humans obtained with a system containing a 1.5 T magnet with a 1 m bore.

Qualitative and Quantitative Aspects of ^{13}C NMR Spectroscopy and Metabolism

In this symposium on stable isotopes in nutrition, it is appropriate to illustrate the nature of the metabolic information obtainable from ^{13}C NMR spectroscopy with an example from metabolism in liver, the liver playing a central role in maintaining the body's nutrients within narrow limits while being presented with a highly variable supply. One of the major regulated processes which occurs in the liver is gluconeogenesis, the resynthesis of glucose from the products of glucose catabolism in brain, muscle, and fat cells. It is important to understand the regulation of this process because the body normally stores only enough glucose to fuel the brain for 12 hours, and thus gluconeogenesis plays a central role in maintaining energy homeostasis. The major metabolic pathways for conversion of these products, which consist of three-carbon skeletons, into glucose which has a six-carbon skeleton, are depicted schematically in Figure 1.

One of the major waste products of glucose metabolism in peripheral tissues is the amino acid alanine, and its use as a substrate for gluconeogenesis requires the disposal of ammonia in the urea cycle. In the following discussion, we will illustrate how the flow of the ^{13}C label from alanine into products, byproducts, and intermediates of gluconeogenesis can be used to obtain qualitative and quantitative information about the relative activities of these pathways.

The ^{13}C NMR spectra summarized in Figure 2 illustrate the flow of ^{13}C label from [3-^{13}C]-L-alanine during gluconeogenesis by a Syrian hamster liver perfused with a Krebs-Henseleit buffer containing the 8 mM-labeled alanine. Although 8 mM represents an unphysiologically high concentration of alanine (normal serum levels are approximately 300 μM), it is necessary to consider the low sensitivity of the NMR technique. In order to achieve gluconeogenic rates sufficient to permit detection of labeled metabolites and intermediates in real time with relatively short accumulation times, a relatively high concentration of alanine was used. The resonances observed in these spectra come from both intra- and extracellular metabolites. However, since the perfused hamster liver almost completely fills the sensitive volume of the Rf coil, resonances from intracellular metabolites dominate in these spectra. Changes in the intensity of resonances from ^{13}C-labeled metabolites show up more clearly in spectra (Figure 3) obtained by subtracting the background spectrum of the liver in the absence of the ^{13}C-labeled substrate. One of the major advantages of ^{13}C NMR as a tool for metabolic studies is that information about the distribution of label at specific carbon sites within metabolites is available directly from the spectrum. The distribution of label in products and intermediates can be used to provide valuable information concerning the metabolic history of the product or intermediate. For example, the early appearance of ^{13}C label in C-2 and C-3 of glutamate and glutamine, but not in C-4 of glutamate and glutamine, provides information about the early activity of the Krebs cycle and the pathway by which the ^{13}C label enters it. If labeled pyruvate (formed by transamination of [3-^{13}C]-L-alanine) enters the Krebs cycle as oxaloacetate through pyruvate carboxylase, then the label will appear in C-2 and C-3 of glutamate as shown in Figure 4. The appearance of label at two sites in glutamate is a

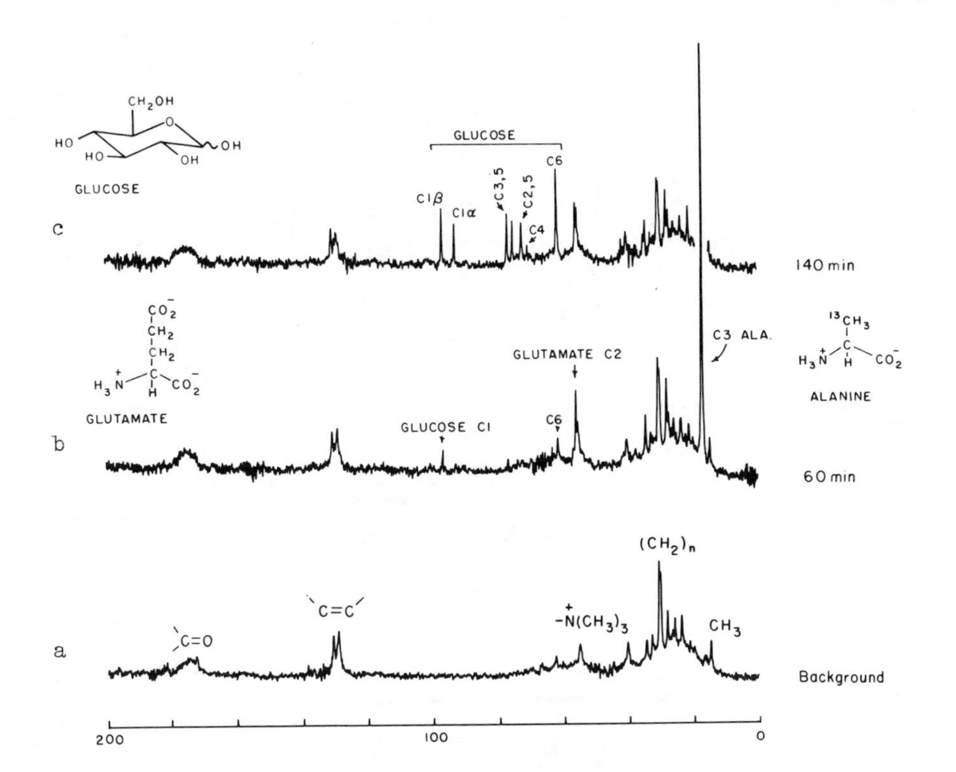

Figure 2. Proton decoupled ^{13}C NMR spectra at 7.0 T from a perfused Syrian hamster liver performing gluconeogenesis from 3-^{13}C alanine. a, Background spectrum before addition of 8 mM alanine; b, spectrum 60 min after addition of alanine; and c, spectrum 120 min after alanine addition. These spectra were accumulated under the following conditions: 37 ^{o}C, 1.9 s pulse internal, 65° pulse angle, 16,000 Hz spectra width, 256 scans, 8.3 min total acquisition time, two level gated broad band decoupling, .4 watt during delay, 7 watt during acquisition.

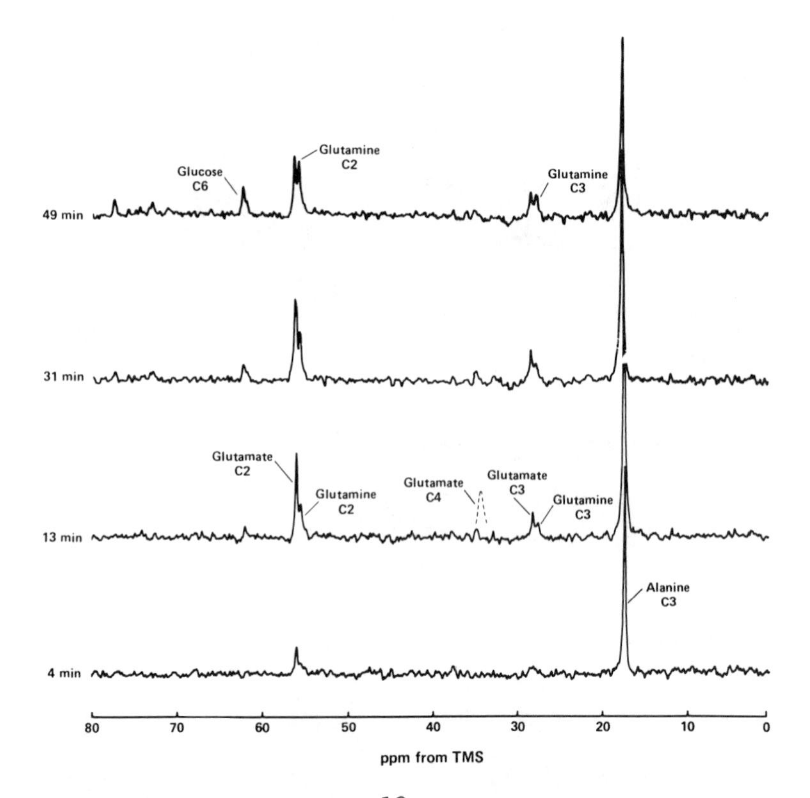

Figure 3. Proton decoupled ^{13}C NMR spectra of perfused liver during first hour of gluconeogenesis obtained by subtracting background liver resonances. The dashed line indicates the chemical shift for C-4 glutamate.

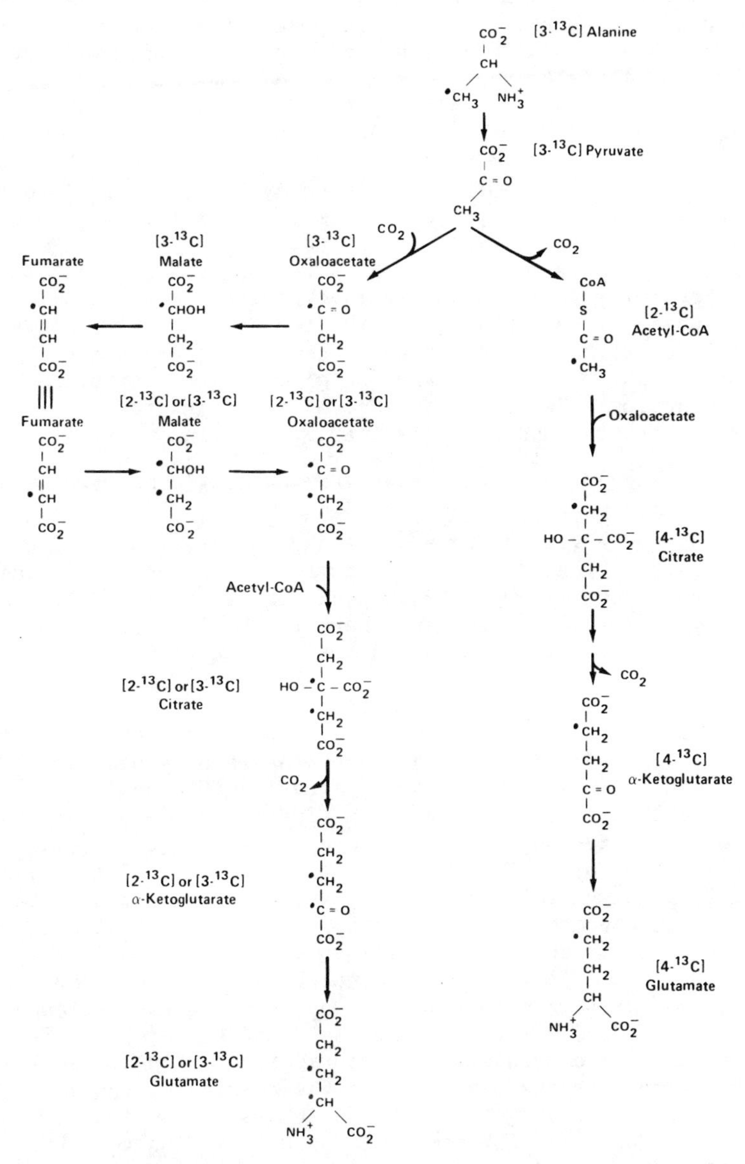

Figure 4. Pathways for glutamate biosynthesis from $3\text{-}^{13}C$ alanine.

consequence of the equilibrium between oxaloacetate, malate and the symmetrical intermediate fumarate, which scrambles the label between C-2 and C-3. Alternately, if pyruvate enters the TCA cycle as [2-^{13}C]-acetylCoA (through pyruvate dehydrogenase and citrate syntase), the label would appear at C-4 of glutamate also as shown in Figure 4.

Since C-2 and C-3 of glutamate are labeled and C-4 is not, all of the [3-^{13}C]-pyruvate derived from [3-^{13}C]-alanine enters the Krebs cycle as oxaloacetate. These results indicate that acetylCoA, which is required for Krebs cycle activity, is derived from an unlabeled pool, probably from the mobilization and oxidation of endogenous lipids in the liver.

Also of interest in Figure 3 is the observation that almost 98% of the label from alanine is incorporated into the amino acids glutamate and glutamine during the first hour of perfusion. This incorporation of label into these amino acids, rather than glucose, during the early stages of gluconeogenesis reflects the requirement for a "metabolically non-toxic" sink for the amino group from alanine and suggests that urea cycle activity in the perfused liver is initially rather low. As shown in later spectra in Figures 2 and 3, the glutamate and glutamine resonances eventually decrease slightly in intensity, and glucose resonances grow. This observation, together with the observation of urea in the perfusate at the end of the experiment, suggests that urea does serve as the major ultimate sink for nitrogen in the liver.

Figures 2 and 3 also show that, in the late stages of gluconeogenesis, all the carbon atoms of glucose are labeled, but to different extents, indicating that more than one metabolic pathway contributes to the formation of glucose. We can use the relative enrichments for the glucose carbons to probe the contributions of various pathways to the synthesis of glucose from labeled alanine. As can be appreciated from Figure 1, there are several metabolic pathways by which alanine can be converted to glucose. In order to interpret the data quantitatively, we have used the method of mixtures. A reasonable subset of the possible metabolic pathways is selected, and the labeling pattern (distribution of isotopomers) for products and/or intermediates is associated with each pathway. The relative enrichment at each site is expressed in terms of the relative fluxes of label through the possible metabolic pathways and the expressions solved to give the relative fluxes as functions of the relative enrichments in the products and intermediates. In our analysis of gluconeogenesis, we have limited the possible pathways to the nine pathways shown together with their associated labeled products in Table I. For completeness we have included the possibility that the acetate pool may be partially labeled (Pathway M).

As shown in Table I, a number of the pathways are degenerate in the sense that they lead to identical labeling patterns. Consequently, our analysis will give only the sum of relative fluxes through degenerate pathways. For convenience in

Table I. Subset of Pathways for Biosynthesis of Glucose and Intermediates During Gluconeogenesis from [3-^{13}C]Alanine

	Pathway	Isotopomer(s)
A	[3-^{13}C]ALA → Pyr → OAA → MAL → OAA → PEP → → →	1,6 Glucose
B	[3-^{13}C]ALA → Pyr → OAA → ASP → PEP → → →	1,6 Glucose
C	[3-^{13}C]ALA → Pyr → OAA → $Fum_{(inside)}$ ⇄ OAA → MAL → → →	1,6;2,5 Glucose
D	[3-^{13}C]ALA → Pyr → OAA → $Fum_{(inside)}$ ⇄ OAA → ASP → → →	1,6;2,5 Glucose
E	[3-^{13}C]ALA → Pyr → OAA → MAL → $Fum_{(outside)}$ ⇄ OAA → PEP → →	1,6;2,5 Glucose
F	[3-^{13}C]ALA → Pyr → OAA → ASP → $Fum_{(outside)}$ ⇄ OAA → PEP → →	1,6;2,5 Glucose
G	[3-^{13}C]ALA → Pyr → OAA → CIT → → → αKG → MAL → OAA → →	3,4 Glucose; CO_2
H	[3-^{13}C]ALA → Pyr → OAA → $Fum_{(inside)}$ ⇄ OAA → CIT → αKG → MAL → →	1,6;2,5;3,4 Glucose; CO_2
M	[3-^{13}C]ALA → Pyr → Acetyl-CoA → αKG → MAL → →	1,6;2,5 Glucose
I	[3-^{13}C]ALA → Pyr → OAA → CIT → → αKG →	2 Glutamate
J	[3-^{13}C]ALA → Pyr → OAA → $Fum_{(inside)}$ ⇄ OAA → CIT → → αKG →	2,3 Glutamate
K	[3-^{13}C]ALA → Pyr → Acetyl-CoA → CIT → → → αKG →	4 Glutamate

Abbreviations - ALA - Alanine, Pyr - Pyruvate, OAA - Oxaloacetate, MAL - Malate, PEP - Phosphoenol Pyruvate, ASP - Aspartate, CIT - Citrate, αKG - αKeto-glutarate

$Fum_{(inside)}$ - Fumarate in equilibrium with the OAA Pool inside the mitochondria

$Fum_{(outside)}$ - Fumarate in equilibrium with the OAA Pool outside the mitochondria

discussing our results in terms of the cellular physiology, we have made the following substitutions: A+B=ST; C+D+E+F=Fum; and G+H=TCA. ST represents the relative flux of label to glucose that is not scrambled by either the intramitochondrial or the cytosolic fumarate pool. Fum is the flux of label that is scrambled by fumarase activity, either in the mitochondria or cytosol. TCA represents the flux of label around the citric acid cycle. For simplicity, we have ignored the possibility that glucose could be derived from intermediates that have passed around the TCA cycle several times. This simplification is expected to result in negligible errors, since very little enrichment at C-1 of glutamate was observed in all experiments.

We have also assumed that the glutamate enrichments are representative of the oxaloacetate pool from which the glucose is derived. This assumption allows the C-2/C-3 and the C-4/(C-2 + C-3) glutamate enrichments to serve as measures of intramitochondrial fumarase scrambling of the oxaloacetate pool and of the enrichment of the acetate pool, respectively.

The expected alteration of the flux through these pathways in response to alteration in nutrient and hormone supply and the redox state of the liver provides us with a useful new probe of the regulatory mechanisms of gluconeogenesis and the health of the liver. Shown in Table II are the relative fluxes of label

Table II. Relative Fluxes Through Pathways of Gluconeogenesis

Substrates	ST	Fum	TCA
8 mM 3-^{13}C Alanine	0.18	0.44	0.38
8 mM 3-^{13}C Alanine + 20 mM Ethanol	0.05	0.71	0.23

from alanine to glucose through the pathways described above, determined from the relative enrichment in the glucose and glutamate present in the perfusate at the end of the perfusions with and without 20 mM ethanol. In order to minimize metabolic variations from livers from different animals, each liver is used as its own control by changing the perfusion medium halfway through the experiment. In addition, spectra of the perfusates were accumulated with gated proton decoupling under fully relaxed pulsing conditions, so that the intensities observed in the spectra accurately reflect the enrichments.

Two trends in this table are noteworthy. The first is the decreased flux of label through the "straight" pathways (ST) in the presence of ethanol. This trend indicates that ethanol markedly increases the scrambling of intermediates by fumarase activity as noted also by Cohen *et al.* (16) in isolated hepatocytes. In addition, ethanol also decreases the relative flux of label through the TCA cycle. This observation suggests that the activity of the TCA cycle is reduced relative to gluconeogenesis in the presence of ethanol, when the redox potential of the liver is increased. The relative activity of the TCA cycle is more strongly controlled by the presence of reduced pyridine nucleotides than by the acetate units produced by ethanol oxidation.

In order to more directly investigate the regulation of metabolism by the redox potential of the cell, we have recently labeled the intracellular pools of pyridine nucleotides in the liver, using the ^{13}C-labeled vitamins, nicotinic acid and nicotinamide.

In the design of these experiments a number of considerations are involved. In order to minimize interferences from background resonances it is desirable to label sites which give resonances in relatively clear regions of the spectrum. From this standpoint, the C-2 carbon represents an attractive labeling site, since the background spectrum of liver is relatively free of interfering resonances at 140 ppm (Table III). However, the chemical shift difference between the oxidized and reduced

Table III. Chemical Shifts of Selected Pyridine Nucleotide Metabolites

	C-2	C-5	C-6
Nicotinate	149.87	124.83	151.17
Nicotinamide	148.29	124.96	152.48
NAD^+	140.72	129.69	143.35
$NADP^+$	140.83	129.70	143.34
NADH	139.15	106.22	125.12
NaMN	142.03	129.34	142.26
NaAD	141.52	129.32	142.39

(abbreviations shown in Figure 6 caption)

resonances is only 1.6 ppm, and chemical exchange (15), or poor resolution may preclude observation of separate reduced and oxidized resonances. The chemical shift at C-5 is more sensitive to the oxidation state, but the oxidized resonance falls in the same region as the intense resonances from the fatty acyl olefinic carbons, making quantitation of the NAD^+ resonance difficult. The relaxation behavior of the carbon site selected is also an important consideration. The selection of protonated carbons has been favored largely because the spin-lattice relaxation times are short, allowing rapid pulse repetition rates and good signal-to-noise ratios in short periods of signal averaging. Protonated carbons offer the additional advantage that their relaxation mechanisms are dominated by a single well-understood interaction, and information about the molecular dynamics of metabolites within cells can be obtained from their relaxation behavior. Narrow linewidths (long spin-spin relaxation times) are also desirable, especially when trying to detect larger metabolites. For the pyridine nucleotides (MW ≈ 700) we have found that methine carbons appear to represent a good compromise, giving narrow resonances with relatively short spin-lattice relaxation times.

A series of ^{13}C spectra showing the incorporation of nicotinate into metabolites of the pyridine salvage pathway in a perfused hamster liver is shown in Figure 5. This series of spectra was accumulated during infusion of 3 mg per hour [2-^{13}C] nicotinic acid into the perfusate. Resonances from the precursor, nicotinate, and two products, nicotinamide and NAD^+ are detected after approximately 1 hour of infusion. Shortly after infusion of the labeled nicotinate ceases (at approximately 255 min), the resonance from the precursor disappears. The appearance and disappearance of these resonances during this experiment are consistent with the biosynthesis of NAD^+ from nicotinate by the Preiss-Handler pathway (Figure 6) [nicotinate → nicotinate mononucleotide (NaMN) → nicotinic acid adenine dinucleotide (NaAD) → nicotinamide adenine dinucleotide (NAD^+)] and the subsequent formation of nicotinamide by glycohydrocase [NAD^+ → nicotinamide]. Also of note in the spectra in Figure 5 is the lack of any detectable resonance from reduced pyridine nucleotides (approximately 1.5 ppm upfield from the NAD^+ resonance).

Figure 7 shows the response of the redox potential in a perfused hamster liver to the addition of 45 mM ethanol. Instead of the _in vitro_ labeling strategy just described, the pyridine nucleotide pools in this hamster liver were labeled _in vivo_ by intraperitoneal injection of 35 mg [5-^{13}C] nicotinamide 5 hours prior to sacrifice. The bottom two spectra (2.6 min and 12.8 min) were obtained prior to addition of ethanol. They show resonances from labeled NAD^+, natural abundance glycogen and natural abundance choline methyl groups of phospholipids but no resonance from reduced pyridine nucleotides. After addition of 45 mM 10% [1-^{13}C] ethanol (at 17.9 min), resonances from C-1 of ethanol and NADH are detectable. These data demonstrate that the pyridine nucleotide pools labeled by intraperitoneal injection are metabolically active and that addition of 45 mM ethanol results in a marked change in the redox potential of the liver as measured by NMR. Furthermore, the observation of separate resonances for the oxidized and reduced pyridine nucleotides indicates that chemical exchange between oxidized and reduced forms is slow on the NMR time scale, and demonstrate that NMR may be used to quantitate the redox potential of free pyridine nucleotides _in situ_.

Since the chemical shifts of carbon nuclei in the nicotinamide moiety of the di-phosphopyridine nucleotides (NAD^+ and NADH) are almost identical to the tri-phosphopyridine nucleotides ($NADP^+$ and NADPH), the intensities of the oxidized (NAD^+) and reduced (NADH) resonances determine a mean reduction charge of the cells defined by

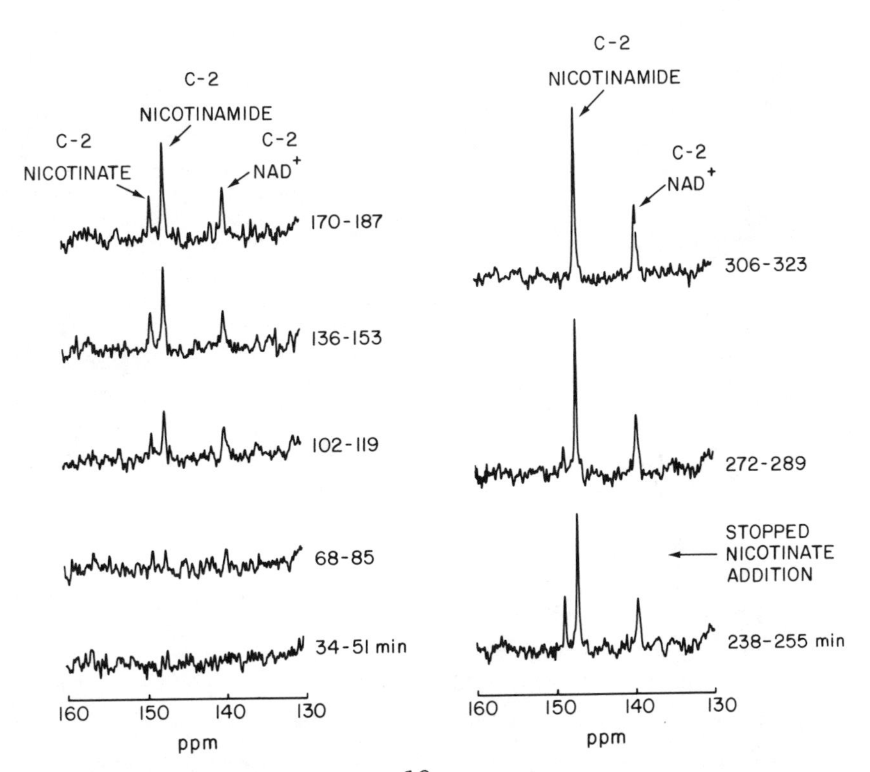

Figure 5. Proton decoupled ^{13}C NMR spectra of perfused liver showing synthesis and degradation of the pyridine nucleotides from 2-^{13}C nicotinate.

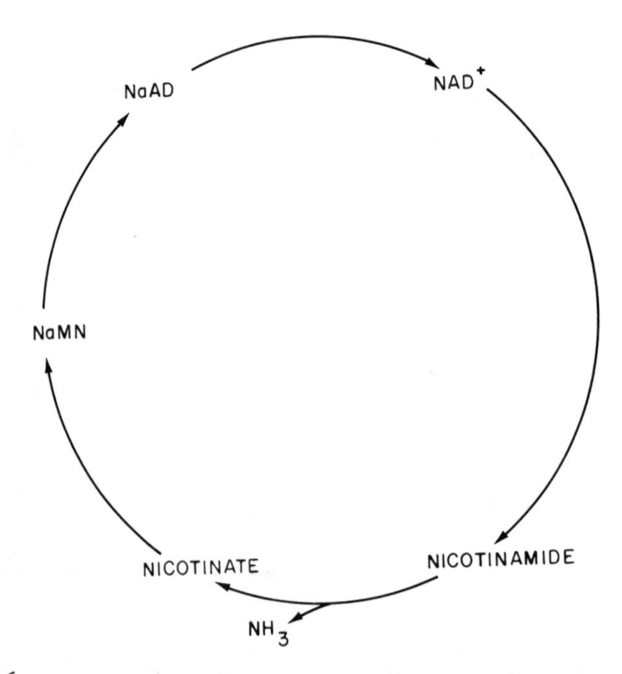

Figure 6. Proposed pyridine nucleotide cycle. Abbreviations: NaMN, nicotinic acid mononucleotide; NaAD, nicotinic acid adenine dinucleotide; and NAD^+, oxidized form of nicotinamide adenine dinucleotide.

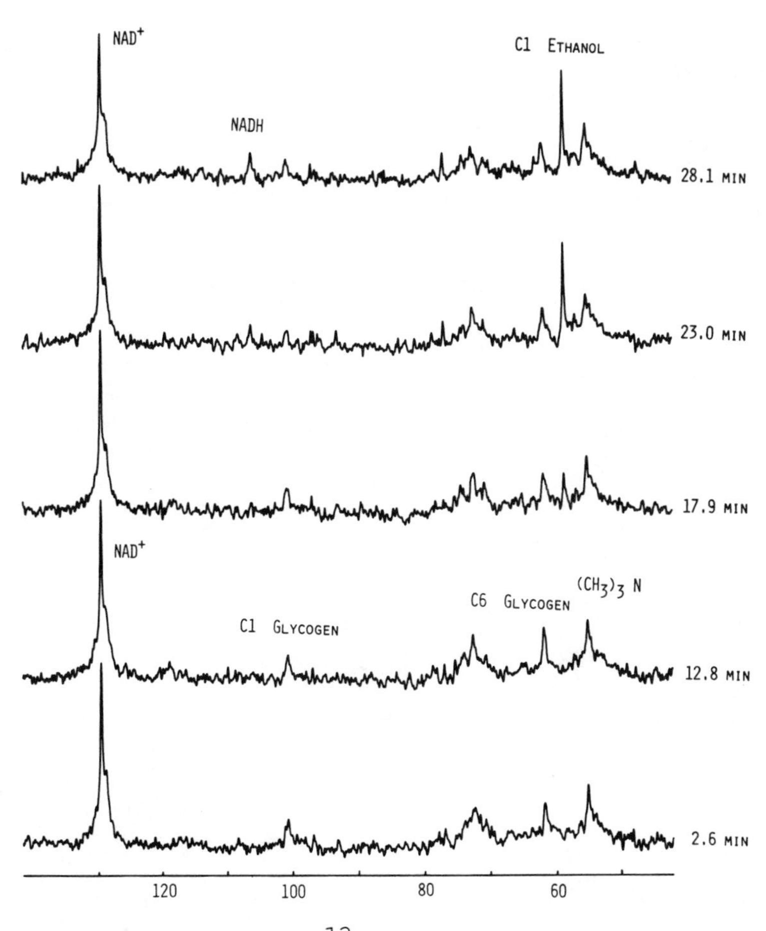

Figure 7. Proton decoupled ^{13}C NMR spectra of perfused liver showing the effect of addition of 45 mM ethanol at 17.9 min on the C-5 resonances from NAD^+ and NADH.

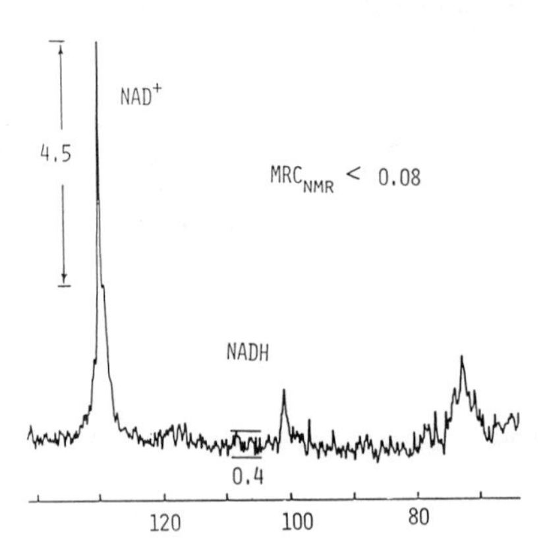

Figure 8. Proton decoupled ^{13}C NMR spectra of perfused liver showing estimated upper limit for the mean reduction charge as observed by ^{13}C NMR.

$$MRC = \frac{[NADH] + [NADPH]}{[NAD^+] + [NADH] + [NADP] + [NADPH]} = \frac{I_{RE}}{I_{RE} + I_{OX}}$$

As noted previously, no resonance from reduced pyridine nucleotides is detectable in well-oxygenated livers, in the absence of reducing substrates. From the signal-to-noise ratio in spectra of well oxygenated liver, an upper limit for the mean reduction charge measured by NMR can be estimated. Figure 8 shows such a spectrum, where the upper limit for the mean reduction charge is approximately 0.08. This estimate is five-fold lower than estimates for the MRC obtained using fluorescence spectroscopy and/or extraction techniques. However, in contrast to these techniques, NMR primarily measures metabolites in the cells which are not bound to enzymes. The differences between the reduction charge in the cell estimated from these *in situ* measurements may reflect the higher affinity that triphosphopyridine nucleotides have for enzymes and suggest that a large fraction of the reduced pyridine nucleotides in the cell may be enzyme bound, and consequently do not give a high resolution NMR signal.

Acknowledgments

This work was done under the auspices of the United States Department of Energy.

Literature Cited

1. Eakin, R. T.; Morgan, L. O.; Gregg, C. T.; Matwiyoff, N. A. *FEBS Lett.* 1972, *28*, 259-64.
2. Den Hollander, J. A.; Brown, T. R.; Ugurbil K.; Shulman, R. G. *Proc. Natl. Acad. Sci.* (USA) 1979, *76*, 6096-6100.
3. Walker, T. E.; Han, C. H.; Kollman, V. H.; London, R. E.; Matwiyoff, N. A. *J. BIOL. Chem.* 1981, *257*, 1189-1195.
4. Matwiyoff, N. A.; London, R. E.; Hutson, J. Y. in "NMR Spectroscopy: New Methods and Applications"; Levy, G. C., Ed.; ACS SYMPOSIUM SERIES No. 191, American Chemical Society: Washington, DC, 1982; pp. 157-186.
5. Iles, R. A.; Stevens, A. N.; Griffiths, J. R. *Progr. Nucl. Mag. Res.* 1982, *15*, 49-200.
6. Hoekenga, D. E.; Brainard, J. R.; Hutson, J. Y.; Matwiyoff, N. A. 2nd Annual Meeting, *Soc. Mag. Res. Medicine*, San Francisco CA, Aug. 1983, ABSTRACT, 159.
7. Ackerman, J. J. H.; Grove, T. H.; Wong, L. G.; Gadiau, D. G.; Radda, G. K. *Nature* 1980, *283*, 167-170.
8. Alger, J. R.; Sillerud, L. O.; Behar, K. L.; Gillies, R. J.; Shulman, R. G.; Gordon. R. E.; Shaw, D.; Hanley, P. E. *Science* 1981, *214*, 660-662.
9. Gordon, R. E.; Hanley, P. E.; Shaw, D. *Progr. Mag. Res.*, 1982, *15*, 1-47.
10. Sillerud, L. O.; Shulman, R. G. *Biochemistry* 1983, *22*, 1087-1094.

11. Alger, J. R.; Behar, K. L.; Rothman, D. L.; Shulman, R. G. 2nd Annual Meeting, Soc. Mag. Res. Medicine, San Francisco, CA, Aug. 1983, ABSTRACTS, p. 1.
12. Neurohr, K.; Barrett, E. J.; Shulman, R. G. Proc. Natl. Acad. Sci. (USA) 1983, 80, 1603-1607.
13. Behar, K. L.; Petroff, O. A. C.; Alger, J. R.; Prichard, J. W.; Shulman, R. G. 2nd Annual Meeting Soc. Mag. Res. Medicine, San Francisco, CA, Aug. 1983, ABSTRACTS, 36-37.
14. Bottemley, P. A.; Smith, L. S.; Edelstein, W. A.; Hart, H. R.; Mueller, O.; Leue, W. M.; Darrow, R.; Redington, R. W. 2nd Annual Meeting Soc. Mag. Res. Medicine, San Francisco, CA, Aug. 1983, ABSTRACTS, 53-54.
15. Coleman, J. E.; Armitage, I. M.; Chlebowski, J. F.; Otvos, J. D.; Schoot Uiterkamp, A. J. M., in "Biological Applications of Magnetic Resonance"; Shulman, R.G., Ed.; Academic: New York, 1970; pp. 345-395.
16. Cohen, S. M., Glynn, P., Shulman, R. G. Proc. Natl. Acad. Sci. (USA) 1981, 78, 60-64.

RECEIVED February 9, 1984

12

Glucose Metabolism in Humans

Studied with Stable Isotopes and Mass Spectrometric Analysis

ROBERT R. WOLFE, MARTA H. WOLFE, and GERALD SHULMAN

Metabolism Unit, Shriners Burns Institute, Departments of Surgery and Anesthesia, University of Texas Medical Branch, Galveston, TX 77550

The primed constant infusion of stable isotopes of glucose is a technique that has enabled many recent advances in the understanding of glucose metabolism in humans. In this chapter we present the theory and validation of the methodology, a description of the analytical procedures involved, and examples of application of the technique. We have tested the response to various rates of exogenous glucose infusion in normal volunteers and hospitalized patients. In normal volunteers, endogenous glucose production is suppressed an amount equivalent to the infused glucose at an infusion ratio as low as 1 mg/kg min, which is less than one-half of the endogenous production rate. In septic and injured patients, however, the suppressibility of glucose production is diminished. In all subjects, between 40 and 50% of the infused glucose was directly oxidized, the exact percent dependent upon the rate and duration of infusion. Glucose labeled with deuterium in various positions can be used to determine the rates of glycolytic/gluconeogenic substrate cycles. Using this approach, we have found that there is a measurable amount of substrate cycling in normal volunteers and that it is significantly depressed in hypothyroid patients.

Over the past several years we have applied stable isotopic tracers to a variety of metabolic and nutritional problems in human subjects. This chapter will focus on examples of our studies of glucose metabolism. The general principle and validation of the tracer techniques we have used will be presented first, followed by a review of some of our work examining the response to infused glucose. In these studies we used stable isotopes because of the

0097-6156/84/0258-0175$06.00/0

health risk of using the analagous radioactive labeled glucose molecules. Finally, we will present results from a recent study in which glucose labeled with deuterium in various positions has been used to quantitate rates of substrate cycling in the gluconeogenic/glycolytic pathway. This later study, although theoretically possible with tritium-labeled glucose, is an example of an experiment in which the use of deuterium was far preferable to tritium, irrespective of the health issue.

Primed Constant Infusion Technique

In a steady-state situation in which there is a constant rate of appearance of a substrate (e.g., glucose) into plasma or extracellular fluid that is matched by an equal rate of tissue uptake, a constant infusion of an isotopically-labeled tracer that is not distinguished metabolically from the unlabeled tracee will cause the enrichment $\frac{[\text{tracer}]}{[\text{tracee + tracer}]}$ of the substrate to rise until an equilibrium (plateau) is achieved. When that equilibrium enrichment is reached, the rate of appearance of the unlabeled tracer can be calculated by dividing the known rate of infusion of isotope by the experimentally determined isotopic enrichment. This general principle was used more than thirty years ago to study plasma glucose kinetics using radiolabeled glucose (1), and remains the backbone of tracer methodology for numerous investigators to this date.

With many substrates of interest, including glucose, the rate of turnover is slow relative to the pool size (concentration x volume of distribution), and consequently several hours of constant tracer infusion may be needed to reach an equilibrium in enrichment. A long infusion time can not only present logistical difficulties, but can also cause interpretive problems. The physiological and metabolic state of the individual may change over the period of infusion, and any recycling of isotope will become accentuated over long infusion times. For these reasons it is useful to give a priming dose with a constant infusion.

The general principle of the primed-constant infusion technique is that if the proper priming dose is given, the sum of the decline in enrichment resulting from the bolus injection and the rise in enrichment due to the continuous infusion equal the ultimate plateau enrichment. If the ideal example is considered in which the concentration of the injected priming dose decays as a single exponential, then the enrichment at any time (t) following the bolus injection is described by the following equation:

$$APE_{(t)} = APE_{(0)}e^{-kt}$$

where k is the rate constant for elimination. If the decay of the injection dose is described by $APE_{(t)} = APE_{(0)}e^{-kt}$, then the $APE_{(t)}$ during a continuous infusion of the same labeled substrate into the same physiological space will be:

$$\text{(iii)}\quad APE_{(t)} = \frac{F}{V \cdot C \cdot k}(1-e^{-kt})$$

where V = volume of distribution, C = concentration of unlabeled substrate, and F = isotope infusion rate.

At plateau: $$APE = \frac{F}{V \cdot C \cdot k}$$

With an appropriate prime in relation to the infusion rate:

$$APE_{(0)}e^{-kt} + \frac{F}{V \cdot C \cdot k}(1-e^{-kt}) = \frac{F}{V \cdot C \cdot k}$$

and: $$\frac{APE_0 \; V \; C}{F} = \frac{1}{k}$$

where APE_0 V C/F = prime/infusion rate ratio.

Since the prime dose (P) = $APE_{(0)}$V C, the P/F ratio is simply the inverse of the rate constant for elimination, which can be determined from a single bolus injection of the isotope. Although the single-exponential curve is often an over simplification of the true kinetics of an injected bolus of isotope, complications in mixing rates, pool size, etc. that influence the nature of decay of isotopic enrichment after a bolus injection will similarly influence the rate at which the enrichment rises during an unprimed constant infusion. Thus, from a practical standpoint, this approach when carefully applied, can shorten the time until an equilibrium is achieved during the constant infusion of labeled glucose from 6-8 hours to 30 minutes or less. From a practical perspective one should wait at least 60 minutes, however, before sampling to ensure attainment of plateau enrichment.

Determination of the oxidation of a substrate by a constant infusion of tracer quantities of a carbon-labeled isotope of the substrate require the determination of the equilibrium enrichment of CO_2 in expired air. In humans this necessitates a constant infusion of the tracer for 7 or 8 hours (2). The necessary asympototic value in CO_2 enrichment can be obtained more rapidly if the bicarbonate pool is primed in addition to the priming of the substrate pool. The rationale for determining the appropriate priming dose for the bicarbonate pool is precisely the same as described above. The additional information that is necessary to calculate the appropriate bicarbonate prime when a primed-constant infusion of a ^{13}C-labeled substrate is given is the percent of substrate uptake that is oxidized to CO_2. In steady-state conditions and at isotopic equilibrium for the plasma substrate being traced, the rate of uptake of tracer equals the rate of infusion of tracer. The rate of production of labeled bicarbonate equals the rate of isotope uptake times the fraction of the uptake directed to oxidation. Calculation of the appropriate bicarbonate prime to give with a primed constant infusion of glucose is then calculated according to the principles outlined above (3).

Validation of the Primed-Constant Infusion Technique We have confirmed experimentally that the theoretical considerations described above apply in the physiological setting (4). Six dogs were anesthetized and all sources of endogenous glucose (liver and kidney) were removed surgically. Glucose was then infused at a known rate, and the constant tracer technique was used to calculate the rate of appearance (Ra) of glucose. The results are shown in Table I. The mean steady-state Ra of unlabeled glucose calculated from the equilibrium enrichment was almost identical to the actual rate of infusion of unlabeled glucose. In the dog study, radioactive glucose was used. However, Figure 1 illustrates the fact that stable isotopes of glucose yield the same turnover data as their radioactive counterparts.

Table I. Ra Calculated by Primed Constant Infusion of Tracer

Dog #	Actual Rate ml/kg min	Tracer Calculated Rate ml/kg min	% Deviation From Acutal Rate
1	2.76	2.82	+ 2.2
2	1.34	1.41	+ 5.2
3	1.54	1.54	0
4	2.05	2.05	+ 0.5
5	2.75	2.60	- 5.5
6	1.87	1.86	- 0.5
7	1.97	1.90	- 3.6
8	1.87	1.87	- 1.1

The coefficient of variation of the tracer calculated rate about actual rate is 3.4%

A limitation in general to the use of a ^{13}C-labeled substrate to measure the rate of oxidation is that the precursor enrichment is diluted by unlabeled substrate as it enters the intracellular compartment before oxidation. With glucose, this is not a problem at rest because plasma glucose is essentially the only source of glucose-intramuscular glycogenolysis only occurs to a significant extent during excercise. Thus, in the resting dog we found the rate of glucose oxidation, as calculated by means of the primed constant infusion of U ^{14}C-glucose, to be most significantly different from the value simultaneously determined by indriect calorimetry (5).

Thus, the primed-constant infusion technique is easily administered, an experiment can be completed in a couple of hours, and valid kinetic data. Because of these attributes, we have applied this technique in a variety of physiological settings.

Glucose Metabolism in Man: Response to Glucose Infusion

Glucose infusion has been used in clinical practice for at least 30 years but the widespread use of large amounts of glucose for nutritional purposes did not become commonplace until the advent of total parenteal nutrition (TPN) in the 1970's. In the early 1970's, glucose was usually the primary caloric source during TPN, and was often given at rates in excess of the level of energy expenditure for the purpose of repleting patients who had lost weight. However, concern over certain side effects associated with high rates of glucose infusion, particularly the development of fatty infiltration of the liver, led to attempts to define optimal glucose infusion rates in different circumstances.

The ultimate nutritional goal of infused glucose is to spare protein. There are two principal mechanisms whereby infused glucose can spare nitrogen. It can result in the suppression of endogenous gluconeogenesis, thereby sparing amino acids for reincorporation into protein, and it can compete with amino acids as a substrate for energy metabolism. In order to investigate these two aspects of the response to glucose infusion in human subjects, we performed a series of experiments using both deuterated and ^{13}C-labeled glucose. The highlights of the ^{13}C glucose method and results are presented below.

Methods

The primed-constant infusion technique was used in all experiments. In order to quantitate both the rate of glucose production and glucose oxidation, it was necessary to determine the enrichment of plasma glucose and of expired CO_2.

All measurements were performed on a gas isotope-ratio mass spectrometer (IRMS) (Nuclide). This is a specialized type of magnetic sector mass spectometer designed specifically to measure isotope ratios of pure gas samples. In the case of CO_2, different collectors allow the precise determination of the ratio of m/e 45 to m/e 44. A dual inlet allows the ratio of the sample gas to be expressed in relation to the ratio of m/e 45 to m/e 44 in a standard gas.

Expired air was collected through a three-way valve into a 5 L anesthesia bag. The contents of the bag were bubbled through 15ml of 0.1 N NaOH to trap the CO_2. Two mls of the resulting Na_2CO_3

solution were pipetted into one arm of a Rittenberg tube and 0.5 ml of 85% H_3PO_4 was pipetted into the other arm. The sample in the tube was then frozen at -86°C and the tube evacuated. The tube was then warmed to room temperature and the contents mixed evolving CO_2 gas. The CO_2 gas was then allowed to enter the IRMS for analysis of $^{13}CO_2$ enrichment.

The enrichment of plasma glucose was determined by first extracting the glucose, combusting it in a vacuum line and then analyzing the enrichment of the resulting CO_2 by IRMS. The extraction was performed by first precipitating the plasma proteins and passing the resultant supernatant sequentially through anion (Dowex AG1-X8) and cation (Dowex AG50W-X9) exchange columns. Most of the water was then evaporated under a stream of N_2 gas, and the sample transferred to a porcelain boat which was subsequently placed in a quartz combustion tube. Some copper filings were also added in the tube to act as a catalyst. The tube was attached to a glass vacuum line (6) which was then evacuated and a combustion oven placed around the quartz tube and heated to 700°C. Oxygen was allowed to circulate through the tube by means of a Toepler pump to ensure complete oxidation. A liquid N_2 trap collected the primary combustion products (CO_2 and H_2O). After complete combustion, the contents of the trap were allowed to warm and the H_2O was separated into a acetone-dry ice slush-trap and the CO_2 was frozen into a sample tube surrounded by liquid N_2 in a dewar flask. The sample tube containing CO_2 was then ready for isotope ratio mass spectrometry as described above.

The ion-exchange chromotography procedure for isolating glucose does not eliminate every possible source of carbon other than glucose from the plasma. The principal contaminant is glycerol. A significant contamination from nonglucose carbons would result in falsely low enrichment values after isotope infusion. Theoretically, glycerol should not be a significant problem since its concentration in plasma is only about 1/100 that of glucose, and there are only three carbons in glycerol. Nontheless, we checked the possibility that there was significant contamination of our glucose sample in two ways.

First, in two dogs we infused U-^{13}C-glucose and withdrew two blood samples (25 cc). We then divided the samples into two aliquots and processed one aliquot as described above. With the other aliquot, the glucose was specifically isolated as potassium gluconate crystals (7) so that there was absolutely no contamination from nonglucose carbons. When the percent enrichment of the carbons from the two aliquots was compared, no significant difference could be detected. We did not use the potassium gluconate technique on all our samples because it requires a large sample and is extremely time consuming.

We further checked the acceptability of our technique for separating glucose for percent enrichment determination by simultaneously infusing both U-^{13}C and U-^{14}C glucose into four dogs and comparing the values obtained for the rate of appearance (Ra) of glucose, calculated by means of both isotopes. If there was a significant contamination of the glucose in which the percent enrichment was measured, the Ra calculated from that data would be higher than when the Ra was calculated from the ^{14}C glucose. Figure 1 shows the data from one experiment; the overall averages showed no significant difference between the two techniques (3).

Results

Regulation of Glucose Production Figure 2 shows the suppressive effects of infused glucose on glucose production in normal, fasting volunteers. In the basal state, the rate of endogenous glucose production is slightly more than 2 mg/kg min. When glucose was infused at 1 mg/kg min there was a precise suppression of glucose production in equivilant amount. It has been documented since the early 1950's that glucose infused at this low rate has a nitrogen sparing effect. The results shown in Figure 2 demonstrate that the entire basis of this N sparing effect is the suppression of gluconeogenesis since the total amount of glucose entering the plasma during the 1 mg/kg min infusion is not changed from the basal state.

High rates of glucose infusion have further suppression effects on the rate of glucose production (Figure 2). When glucose is infused at the rate of 4 mg/kg min, glucose production from nonrecycled glucose carbons is completely suppressed in normal volunteers (Figure 2). The same holds true in non-stressed hospitalized patients. In the situation of severe injury (8) or sepsis (9), endogenous gluconeogenesis persists during a 4 mg/kg min infusion but can be suppressed with higher rates of glucose infusion. In all physiological settings a point is reached at which further increases in the rate of glucose infusion cannot induce any greater N sparing as a result of suppression of gluconeogenesis. In all situations we have tested to date, this point is below caloric requirments. Nutritional benefit from glucose infusions at rates in excess of the amount needed to maximally suppress glucose production must therefore be justified on the basis of the direct oxidation of the infused glucose.

Figure 3 shows the results of an experiment in which hospitalized patients were infused at progressively increasing glucose infusion rates over a six day period (10). Glucose production was suppressed completely at the lowest infusion rate tested (4 mg/kg min). During the first two hours of the 4 mg/kg min infusion, only about 40% of the infused glucose was directly oxidized to CO_2. After two days, there was an adaptation that resulted in an increase in the percent of infused glucose directly oxidized, but it still only reached 50%. As the glucose infusion rates were increaed over the next four days, there were minimal increases in glucose oxidation (Figure3). When expressed as the % of total CO_2 production derived from the direct

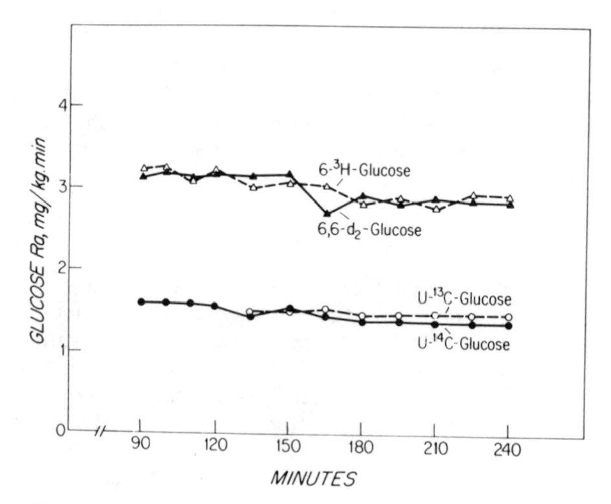

Figure 1. Representative example of experiments in dogs showing that comparable data are obtained when radio-labeled glucose and the analogs stable isotope of glucose is used to determine the rate of glucose production (Ra).

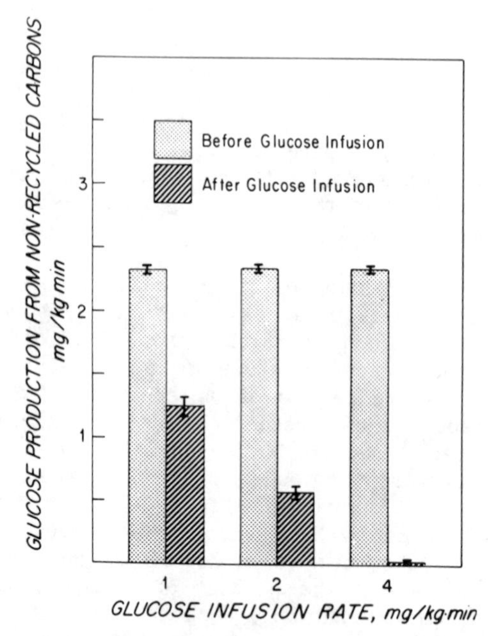

Figure 2. Effect of exogenous glucose infusion on endogenous glucose production in normal volunteers.

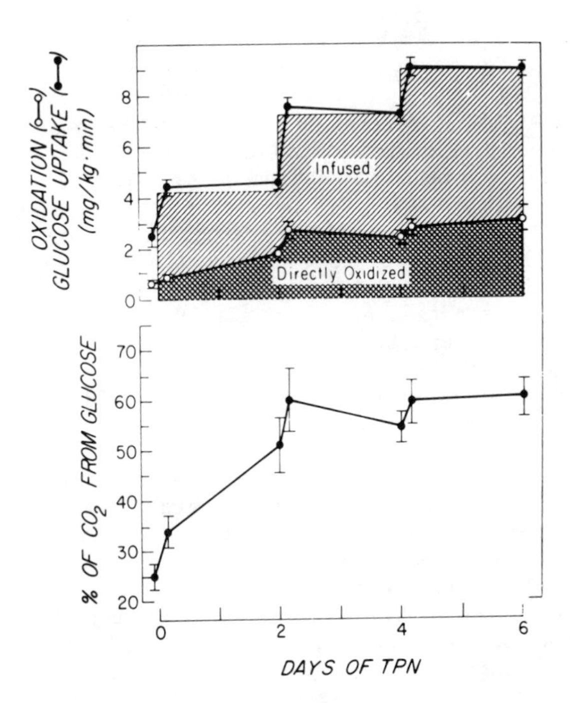

Figure 3. Glucose oxidation measured by means of U-^{13}C-glucose in hospitalized patients receiving total parenteral nutrition.

oxidation of glucose, a maximum value of 65% was achieved even when glucose was infused at the rate of (9 mg/kg min) that provided greater than the entire caloric requirement of the individual.

These data led to the conclusion that there is a limit to the amount of glucose that can be directly oxidized. Additional glucose is stored as glycogen and, when those stores are filled, is converted to fat in the liver. That fat is in turn transported to periphheral tissues where it might be either oxidized or stored. One consequence of the conversion of glucose to fat at a high rate is the production of excess CO_2 (10), since the respiratory quotient of fat synthesis is 8.7. This extra CO_2 load can be a problem in patients in whom respiratory function is already impaired (11). A second side effect of the high-dose glucose is the deposition of fat in the liver, since only a small fraction of the newly synthesized fat needs to remain in the liver to cause such a problem. Because of these side effects of overdoses of glucose, it is desirable to determine the optimal glucose infusion rate under different clinical situations, and then not exceed that rate. Using the approach outlined here we have been able to estimate the appropriate glucose infusion rate in a variety of clinical situations (e.g., 8,9,10).

Substrate Cycling

A substrate cycle is produced when a non-equilibrium reaction in the forward direction of a pathway is opposed by another non-equilibrium reaction in the reverse direction of the pathway. The two opposing reactions must be catalyzed by separate enzymes. Such a cycle is called "futile" if both enzymes are simultaneously active, allowing substrate 1 to be converted to substrate 2 and substrate 2 converted to substrate 1 with the net results being chemical energy being converted to heat. Three examples of such cycles exist in the metabolic pathways involving glycolysis and gluconeogensis and are shown in Figure 4. Until the 1970's, it was generally thought that the glycolytic enzymes were completely suppressed during active gluconeogenesis. However, over the past several years the existence of such cycles have been demonstrated to occur *in vitro*. There are few *in vitro* studies that demonstrate the existence of substrate cycles, and furthermore, their existence in man has yet to be shown.

A role for substrate cycling in the provision of sensitivity and flexibility in metabolic regulations was first proposed by Newsholme (12). An increase in sensitivity is achieved, for example, in phosphofructokinase, since at rest the net flux through phosphofructokinase may be reduced to zero as a result of the biophosphatase reaction. A relatively small rise in one of the enzymes above the other will result in a marked increase in net flux. Furthermore, the higher the rate of cycling as compared to the net flux in the result state, the greater the sensitivity (12). It is also possible that substrate cycles could play an important role in thermogenesis and thus be of importance in weight control (13).

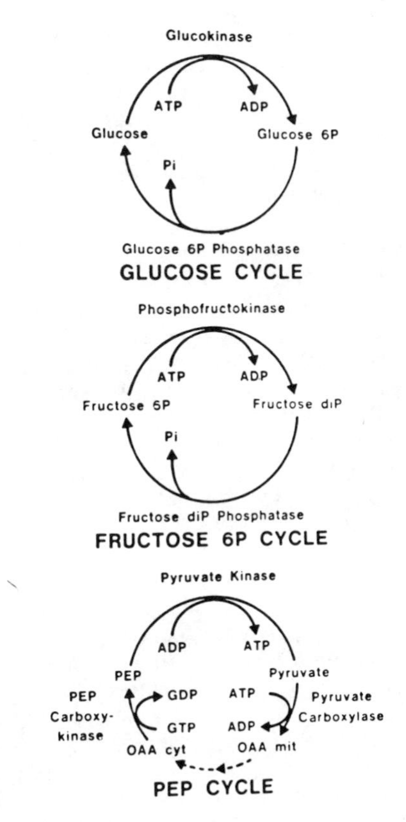

Figure 4. Examples of substrate cycles in the gluconeogenic/glycolytic pathway.

It is possible to quantitate the rate of flux through the cycles illustrated in Figure 4 by using glucose labeled in different positions. The principle is that the rate of flux through a cycle can be calculated by the difference in flux rates obtained when two differently labeled molecules of glucose are used. One of the labels must be lost when the molecule passes through the cycle. The other label must not be lost when passing through the cycle of interest, but must be lost on passage through the next cycle. We have used this approach to measure the flux through cycles 1 and 2, illustrated in Figure 4. We have accomplished this by simultaneously infusing 2-d-glucose, 3-d-glucose, and 6,6-d_2-glucose. Deuterium in the 2 position is lost in the hexose-isomerase reaction (glucose-6-p <---> fructose-6-p). Due to the high activity of this isomerase, most of the d-2 will be removed before glucose-6-p leaves the pool either for glycogen or for plasma glucose.

Because 2-d is lost after passage through the first substrate cycle, use of that tracer will yield the value for total glucose production plus substrate cycles 1,2, and 3. (3-d) glucose can cycle through substrate cycle 1 and not lose its label, but will lose its label if it passes through cycle 2 because in the aldolase cleavage of FDP, the d-3 appears on the C-1 of dihyroxyacetone phosphate. In the isomeration to glyceraldehyde-3-phosphate, 3-d is exchanged with protons of water. Thus, Ra as determined with 3-d-glucose will include glucose production plus substrate cycles 2 and 3. Ra measured by means of 6,6-d_2-glucose will not include recyclling through cycles 1 and 2, but will include recycling through the pyruvate-phosphoenopyruvate-pyruvate cycle (cycle 3). This is because one deuterium in the six positions is lost at the pyruvate carboxylase reaction and the other is lost in the equilibration with hydrogen pool of the mitochondria during the equilibration between oxaloacetate, malate and fumarate.

Calculation of the rates of cycling through cycles 1 and 2 is then calculated by subtraction. We have currently used this approach to quantitate substrate cycling in normal volunteers and have assessed the effect of the thyroid hormones on the rates of substrate cycling by studying hypothyroid patients and hyper-thyroid patients.

<u>Analysis</u> All analyses were performed on a Hewlett-Packard 5985 B quadrupole gas chomatograph mass spectrometer. Plasma glucose was initially separated using the technique described above and the penta-acetate derivative of glucose was formed by adding acetic anhydride and pyridine to the dried glucose sample and heating at 100 C for five minutes. Approximately 1ul of the acetic anhydride solution was required for analysis.

A three foot glass coil packed with 3% OV101 was used for chromotographic separation with temperature programmed from 175°-250° at 20° per minute. In the electron impact (EI) mode, He carrier gas was used with a flow rate of 20 ml/min. In the chemical ionization (CI) mode methane was used both as the carrier and the reagent gas with a flow rate of 20 ml/min. The retention time was approximately 2 minutes. Two anomeric peaks may be observed and can be integrated together or separately.

Since the samples contained three isotopic isomers with potential variation in enrichment at each position, all three positions needed to be measured separately for each data point. However, because of the nature of the structure of the molecule and the resultant fragmentation patterns, none of the three positions could be measured without interference from one or both of the other two. The use of other derivatives, such as the butyl-boronate or tri- methylsilyl derivatives, did not improve this situation. Thus, corrections for the interfering enrichments needed to be made. Both (70eV) electron impact ionization (EI) and methane chemical ionization (CI) were used to take advantage of particular fragments or enhanced signal abundances which occurred in each mode.

The sum of the $2d_1$ and $3d_1$ enrichments were measured by monitoring the ratio of the ion abundance at m/e 116.1 to that at m/e 115.1 in the EI spectrum. The ionic fragment at mass 115.1 contains both the 2 and 3 carbon of glucose. However, it does not contain the number 6 carbon. This was verified by analyzing the fragmentation patterns of the three infusates individually.

The ion in the chemical ionization spectrum occurring at m/e 169.1 contains the 3 and the 6 carbon, but does not contain the 2 carbon. Therefore the $3d_1$ (m+1) value was measured by monitoring the ratio of m/e 170.1 to 169.1 and the $6d_2$ (m+2) value was measured by monitoring the m/e ratio of 171.1 to 169.1. A correction factor must be applied for the interfering signal at m/e 171.1 created by the $3d_1$ spectrum overlapping the 171.1 abundance of the $6d_2$ spectrum. Similarly, the 170.1 value in the $3d_1$ measurement must be corrected for the 170.1 overlap from the $6d_2$ spectrum. After determining the corrected value for the enrichment of the $3d_1$ molecule, this value was subtracted from the measured value at the m/e 116.1 to obtain the $2d_1$ enrichment.

The atom percent excess (APE) of the labeled molecules is defined as the enrichment of the isotopic ion relative to the sum of all isotopic and non-isotopic ions at the same mass.

APE = [R/(R+1)] 100% where R is the ratio difference between the enriched and background value normalized to 1.

$$= \frac{(m+x)_s\% - (m+x)_b\%}{(m+x)_s\% - (m+x)_b\% + 1} \quad (100\%)$$

where (m+x) refers to the abundance of the isotopic ion relative to the non-isotopic ion (m) normalized to 100%, for the sample (s) and background (b).

The correction factors were obtained empirically with labeled standards.

Results

The results are shown in Table II. The results clearly demonstrate the existence of substrate cycles in normal human volunteers.

Table II. Rates of Appearance of Glucose Calculated with Different Tracers[1]

	Ra(2-d)	Rd(3-d)	Rd(6,6-d)	Cycle 1	Cycle 2
Normal Volunteers	3.23±0.56	2.64±0.49	2.00±0.27	0.59±0.73	0.63±0.56
Hyper-thyroid	3.92±0.20	3.51±0.57	2.55±0.08	0.81±0.84	0.97±0.48
Hypo-thyroid*	1.77±0.56	1.52±0.37	1.57±0.31	0.24±0.26	

[1]Units are mg/kg min. *In some subjects only 2 and 6,6-d_2 were used. Therefore the recycling figure is the total of both cycles.

Secondly, the role of thyroid hormones in stimulating substrate cycling is suggested by the significant reduction in cycling in hypothyroid patients. This reduction occurred in conjunction with a fall in metabolic rate, but the difference in heat production that could be attributed to the substrate cycling was small in relation to the difference in metabolic rate. The lack of a causal relationship between thyroid-induced substrate cycling and metabolic rate is further suggested by the fact that although the metabolic rate was significantly elevated in the hyperthyroid patients, the elevation in the rate of substrate cycling failed to reach statistical significance in the hyperthryroid patients.

The failure to explain changes in metabolic rate by changes in the flux through these substrate cycles does not negate the possible significance of such cycles throughout the body. The important parts of our work on this topic to date are that we can use stable isotopes to quantitate substrate cycling in humans, and that these cycles are under hormonal control to some extent. Further studies using these techniques should clarify in more detail the factors that control the rate of flux through substrate cycles and their physiological significance.

Literature Cited

1. Stetten, D. W.; Welt, E. D.; Ingle, D. I.; Morley, E. H. J. Biol. Chem. 1951, 192, 817-830.
2. Issekutz, B., Jr.; Paul, P.; Miller, H. I.; Bortz, W. M. Metabolism 1968, 17, 62-73.
3. Wolfe, R. R.; Allsop, J. R.; Burke, J. R. Metabolism 1979, 28, 210-220.
4. Allsop, J. R.; Wolfe, R. R. Biochem. J. 1978, 172, 407-416.
5. Wolfe, R. R.; Durkot, M. J.; Wolfe, M. H. Amer. J. Physiol. 1981, 241, E385-E395.
6. Craig, H. Geochimica. et. Cosmochimica. Acta. 1953, 3, 53-92.
7. Blair, A.; Segal, S. J. Lab. Clin. Med. 1960, 55, 955-964.
8. Wolfe, R. R.; Durkot, M. J.; Allsop, J. R. Metabolism 1979, 28, 1031-1039.
9. Shaw, J. H. F.; Wolfe, R. R. Surgery, (In Press).
10. Wolfe, R. R.; O'Donnell, T. F., Jr.; Stone, M. D.; Richmond, D.A. Metabolism, 1980, 19, 892-900.
11. Askanazi, J.; Nordenstrom, J.; Rosenbaum, S.H.; Kinney, J. Anesthesiology, 1981, 54, 373-377.
12. Newsholme, E. A.; Crabtree, J. Biochem. Soc. Symp. 1976, 41, 61-119.
13. Dunn, A.; Katz, J.; Golden, S.; Chenoweth, M. Am. J. Physiol. 1976, 230, 138-142.

RECEIVED January 31, 1984

Stable Carbon Isotope Ratios as Indicators of Prehistoric Human Diet

T. W. BOUTTON and P. D. KLEIN—Stable Isotope Program, U.S. Department of Agriculture, Agricultural Research Service, Children's Nutrition Research Center, Department of Pediatrics, Baylor College of Medicine, Texas Children's Hospital, Houston, TX 77030

M. J. LYNOTT—U.S. Department of the Interior, National Park Service, Federal Building Room 474, Lincoln, NE 68508

J. E. PRICE—Center for Archaeological Research, Southwest Missouri State University, Naylor, MO 63953

L. L. TIESZEN—Department of Biology, Augustana College, Sioux Falls, SD 57197

Stable carbon isotope analyses of bone collagen extracted from prehistoric human skeletal remains from southeastern Missouri and northeastern Arkansas indicate that intensive maize agriculture began in this region around 1000 AD, that the incorporation of maize as a significant component of the human diet was rapid, and that 35 to 72% of the human diet from 1000 to 1600 AD consisted of maize. While archaeologists generally have thought that maize agriculture was common by the start of the Mississippian culture ($\simeq$ 700 AD), these results indicate that maize agriculture did not reach full development until some time after 1000 AD. Stable carbon isotope methods exploit natural variations in the relative abundances of the isotopes ^{13}C and ^{12}C, and are valuable archaeological tools for the study of paleonutrition.

The use of the stable isotope ^{13}C as a tracer in biological research has become increasingly common as evidenced by recent bibliographies (1,2). The effective use of this isotope has been established in the field of nutrition, where it has been applied in human clinical studies (3,4), in food science research (5), and in ecological studies of animal food habits (6,7).

Many nutrition studies are carried out at natural abundance levels of ^{13}C. Because these levels are low and because differences in the ^{13}C content of natural materials are small, stable carbon isotope ratios ($^{13}C/^{12}C$) are expressed in relative terms as $\delta^{13}C$ values. A $\delta^{13}C$ value represents the per mil (parts per thousand) deviation of the ^{13}C content of the sample from the international PDB limestone standard, the $\delta^{13}C$ value of which has been set arbitrarily to 0 ‰. Thus, a $\delta^{13}C$ value of -27.0 ‰ would mean that the sample contained 27 parts per thousand less ^{13}C than the PDB standard. Although the PDB standard no

0097–6156/84/0258–0191$06.00/0

longer exists, the National Bureau of Standards distributes several reference materials which can be related back to PDB.

The reason for the use of natural abundance levels of ^{13}C in such a wide variety of nutritional studies is that there are small but predictable differences in the $^{13}C/^{12}C$ content of natural materials (Figure 1). All naturally occurring reduced organic carbon is depleted in ^{13}C relative to PDB, carbonates, and atmospheric CO_2 (8-10). The source of much of the variation in the ^{13}C content of organic carbon ultimately can be traced back to the process of photosynthesis.

Carbon Isotope Fractionation in Nature

Most plants reduce CO_2 to carbohydrate according to the well-known Calvin-Benson or C_3 pathway, where the initial product of photosynthesis is the 3C compound phosphoglycerate. Fixation of CO_2 to phosphoglycerate occurs with the assistance of the enzyme ribulose bisphosphate (RuBP) carboxylase, which discriminates heavily against $^{13}CO_2$ (11). Consequently, plants with C_3 photosynthesis have $\delta^{13}C$ values that average -27.0 ‰ (12). Plants with the Hatch-Slack or C_4 photosynthetic pathway initially fix CO_2 into the 4C malic or aspartic acids, catalyzed by the enzyme phosphoenolpyruvate (PEP) carboxylase. This enzyme discriminates much less against $^{13}CO_2$ (13), so that C_4 plants have $\delta^{13}C$ values closer to that of atmospheric CO_2, averaging -12.5 ‰ (12). Due to the physiological characteristics that result from the C_4 pathway, C_4 plants are most common in warm environments where water may be limiting, such as grasslands, deserts, and salt marshes. Examples of common C_4 plants include corn, sugar cane, sorghum, and many grasses important in grazing land in the southern and western United States.

A few species of plants are capable of reducing carbon via either the C_3 or C_4 pathway and are known as Crassulacean acid metabolism or CAM plants. As a result of their photosynthetic flexibility, their $\delta^{13}C$ values most commonly range from approximately -12 to -27 ‰ (14). CAM plants are succulents, such as the cacti, and are seldom abundant or of economic importance.

In addition to the fact that plants differ in their $\delta^{13}C$ values, another factor of major importance in nutritional studies is that the carbon isotope ratios of animal tissues and products (e.g., feces or breath CO_2) resemble the isotopic composition of animal diets (15-18). Consequently, by analyzing animal tissue, feces, or stomach contents, it is possible to determine whether an animal's diet consists of C_3 plants, C_4 plants, or a mixture of both. In ecological studies, the technique is most useful in situations where C_3 and C_4 plants coexist, such as in grasslands or deserts.

Stable Carbon Isotope Ratios as Archaeological Tools

An area of human nutrition that promises to benefit from stable carbon isotope techniques is paleonutrition, or the study of past diets. Until recently, methods of studying diets of prehistoric people were based largely on the recovery of fragile and poorly preserved plant and animal remains. Dietary information based on these methods is fragmentary and often biased by differential preservation of food items. For example, animal bones left from a human meal would tend to be more resistant to decomposition over time than plant remains, and thus would be overestimated in importance in any attempt to reconstruct a diet. At best, dietary reconstructions based on food remains can indicate only what items may have been eaten, and yield only indirect information concerning the relative importance of the food items.

Human bones frequently are well-preserved components at archaeological sites, and in many cases retain chemical constituents without exchange with the environment that contain information about diet (19-21). For example, ratios of strontium to calcium in bones indicate the relative importance of meat vs vegetable material in a diet.

Among the most important questions in paleonutrition are 1) when did agriculture become an important part of man's subsistence program, and 2) what was the relative importance of cultivated vs native plants? Stable carbon isotope analysis offers a technique for recognition of maize, or corn, in the diets of prehistoric people who lived in eastern portions of the United States. Until recent times, the eastern United States was covered with deciduous forest containing only plants with the C_3 photosynthetic pathway (22,23). Thus, all the plants and the animals dependent on them, including humans, must have been relatively depleted in ^{13}C. Corn, however, a grass with tropical origins, has the C_4 photosynthetic pathway, and is consequently enriched in ^{13}C. Therefore, the introduction of corn into the diets of people living in areas dominated by C_3 plants should be detectable as an increase in the ^{13}C content of body tissues, including bone.

To supplement ongoing archeological studies in southeastern Missouri and northeastern Arkansas, we undertook a study of the ^{13}C content of prehistoric human bones to determine when corn became a component of the human diet in this area, and to estimate the relative importance of this food item in the diet after it was introduced.

Methods

Preparation of Bone Samples. Bone samples from 20 individuals dating from 3200 BC to 1880 AD were obtained from 14 archaeological sites from several archaeologists who have worked in northeastern Arkansas and southeastern Missouri. Samples included skeletal remains from sites in the Mississippi River alluvial valley and the eastern highlands of the Ozark Mountains (Figure 2). It was not possible to analyze bone from a single skeletal position; therefore, bone fragments came from a variety of locations within the skeleton. The ages of the bones had been determined previously by radiocarbon dating. Sex, social status, and age at death were not known for the samples used in this study.

Not all portions of bone are equally suitable for $\delta^{13}C$ analysis of diet. Carbonate carbon is plentiful in bone tissue, but is known to undergo exchange with carbonates in the environment. Therefore, bone carbon

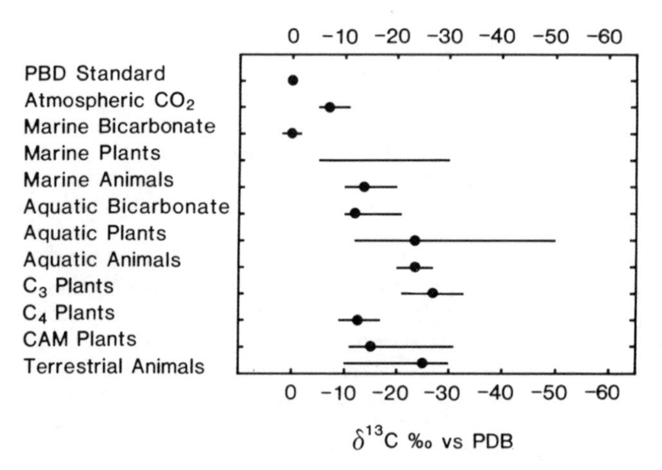

Figure 1. $\delta^{13}C$ values of carbon in some natural materials (8-10).

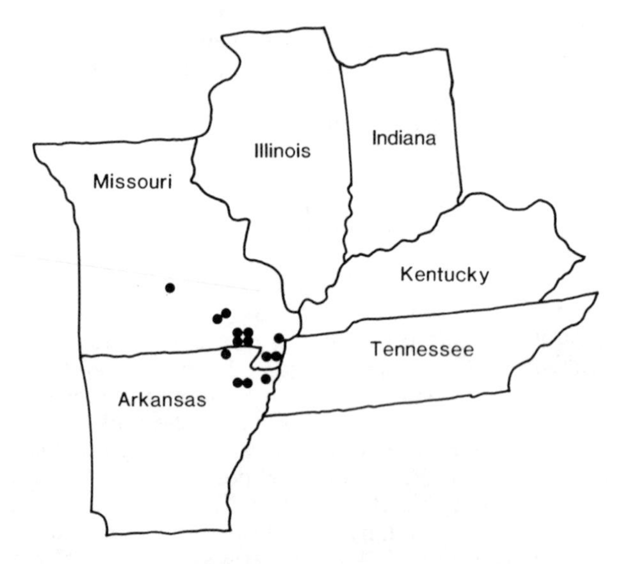

Figure 2. Dots indicate locations of archaeological sites in SE Missouri and NE Arkansas.

present in carbonates is unsuitable both for radiocarbon dating and dietary analysis (24-26). Organic carbon in bone is found largely in the structural protein, collagen, which does not suffer from exchange phenomena (24, 27) and under ideal conditions may be preserved for several million years. For the above reasons and because collagen carbon can originate only from dietary carbon, this protein is the logical choice for dietary analysis of bone using $^{13}C/^{12}C$ ratios.

Bones recovered from archaeological contexts frequently are contaminated with humic substances which are base-soluble, organic products of plant decomposition. Humic substances isolated from soils beneath C_3 plant communities have $\delta^{13}C$ values ranging from -22 to -25 ‰ (28). These contaminants are removed most frequently by treating the bone with NaOH. However, collagen is slightly soluble in salt solutions (29). Treatment of fresh bone, unaffected by contaminants, with NaOH results in a ^{13}C enrichment of collagen of 1 ‰ more than the collagen of bone that has not been treated (30). So, despite the fact that some collagen is removed from the bone during NaOH treatment, there is only a small predictable effect on the $\delta^{13}C$ value of the collagen subsequently isolated.

Soil was removed manually from bone fragments followed by sonication in distilled water for 15 min. Samples were oven-dried at 50°C and then ground in a Wiley mill to pass through a 20-mesh screen. Exchangeable carbonates were removed from the pulverized samples by placing 2 to 5 g of sample in 200 ml of 1 M HCl at room temperature for 24 hours, and stirring several times during this period. The HCl then was discarded, and the pulverized bone washed to neutrality with distilled water. To remove soil humic substances, the sample was placed in 1 M NaOH at room temperature for 24 hours and agitated frequently. Humic substances were recovered from the NaOH by evaporation and saved for isotopic analysis, and the bone sample was washed to neutrality with distilled water. Collagen was solubilized from the bone by incubation in distilled water at pH = 3 at 90°C for 24 hours and stirred occasionally (30-32). The mixture was centrifuged at 3000 x g for 15 min to precipitate the bone fragments. The supernatant was removed and dried at 60°C to obtain the collagen, and the pellet was discarded. To test the reproducibility of the extraction procedure, several bone samples were subdivided and replicate extractions made.

To verify that this method did remove collagen from the bone samples, the extract was subjected to amino acid analysis. Five collagen samples ranging in age from 1000 BC to 1880 AD were dried to constant weight and then hydrolyzed in 6 N HCl for 24 hours at 110°C. Approximately 0.10 to 0.15 g of the dried hydrolysate was dissolved in 3 ml of sodium citrate buffer, and a 50 µl aliquot was analyzed on a Beckman 121 MB amino acid analyzer.

Organic substances such as collagen must be converted to CO_2 for high precision determination of stable carbon isotopic abundances. Solid organic samples are converted to CO_2 by combustion in the presence of an oxidizing agent (33, 34). Quartz tubes (9 mm) were cut to 30 cm lengths, sealed at one end, and loaded with 2 g of CuO and a 9 mm^2 piece of silver foil. The combustion tubes loaded with oxidant then were heated in a muffle furnace at 850°C for one hour to bake out potential organic contaminants. After cooling, 5 to 10 mg of collagen were mixed with the CuO. Sample tubes were attached to a vacuum manifold, evacuated to 10^{-2} torr, and sealed with a torch. Sealed tubes were combusted in a muffle furnace at 850°C for one hour and then cooled to room temperature.

Mass Spectrometric Analysis. Combusted sample tubes were attached to a purification vacuum line connected to the inlet system of the mass spectrometer. Sample tubes were opened under vacuum using a tube-cracker (35), and the gases were passed through a dry ice trap to remove water vapor and a liquid nitrogen trap to collect CO_2. Noncondensible gases were pumped away. The purified CO_2 was thawed and admitted into the inlet system of the mass spectrometer for determination of isotopic composition.

Isotope ratios were measured on a Micromass 602E, a dual inlet, double collector mass spectrometer. The mass 45 to mass 44 ratio of CO_2 from the sample material was compared with that of a standard gas of known isotopic composition. Results are expressed as:

$$\delta^{13}C \text{ ‰} = \left[\frac{R_{sample} - R_{standard}}{R_{standard}}\right] \times 10^3 \quad (1)$$

where R is the mass 45 to mass 44 ratio. Values were corrected for errors from switching valve leakage, ^{17}O contribution to mass 45 abundance, peak tailing, and zero enrichment. All results are reported relative to CO_2 from carbonate from the international PDB standard (8,36).

Mass spectrometer precision was determined by making repeated measurements on a gas sample prepared by combustion of carbon isotope reference material NBS-22. The standard deviation of the mean derived from 10 consecutive measurements of this gas was 0.02 ‰. The error associated with the combustion and purification procedure was measured by replicate combustions of the NBS-22 reference material, which resulted in a standard deviation of 0.12 ‰ for five samples. Thus, the overall precision associated with the mass spectrometric measurement of $\delta^{13}C$ vs PDB was 0.12 ‰, or in absolute terms, 1.3 ppm. Most of the error clearly was associated with the combustion and sample handling process. Since sealed-tube combustions have been shown to produce theoretical recoveries of carbon (33), these small errors most likely arise from handling the CO_2 after it is released from the sample tube.

Results

The amino acid composition of the material recovered from the extraction procedure is representative of known collagen samples from contemporary humans (Table I). Both the amino acid composition and the relative abundances of the individual amino acids are characteristic of

Table I. Amino acid composition of human bone collagen

Amino Acid	1000 BC 23BU26	200 AD 3PO467	500 AD Christensen	1600 AD 23PM5	1880 AD Euro-American	Mean
Glycine	824*	2114	3582	1507	2790	2163
Alanine	240	642	2756	360	1076	1015
Proline	240	720	1125	500	971	711
Aspartic acid	148	479	729	324	616	459
Glutamic acid	151	450	691	307	605	441
Arginine	86	247	417	178	364	258
Serine	66	168	329	130	272	193
Lysine	66	171	276	119	232	173
Leucine	51	145	241	104	213	151
Valine	91	150	197	74	151	133
Threonine	34	105	162	70	140	102
Phenylalanine	34	67	109	31	87	66
Isoleucine	20	58	93	41	78	58
Histidine	9	20	46	20	48	29
Methionine	6	12	51	16	45	26
Tyrosine	3	--	23	--	25	17

*μ moles/g

human bone collagen (37). Hydroxyproline and hydroxylysine, amino acids unique to collagen and some plant proteins, could not be measured by the methods employed. Despite the absence of data on these characteristic amino acids, no other known proteins have an amino acid composition with the relative abundances found in Table II.

$\delta^{13}C$ values of the material solubilized during NaOH treatment were intermediate between those of the collagen and humic material, which suggests that both collagen and humic substances were solubilized in the NaOH. Based on simple isotopic mass balance where collagen and humic acids are sources, as much as 25% of the NaOH extract may consist of collagen.

The combusted collagen samples ranged from 20 to 45% carbon by weight. Since collagen consists of approximately 45% carbon, some of the samples contained inorganic contaminants, probably salts and minerals adhering to collagen fibrils. Despite this evidence of inorganic contamination, all standard deviations of $\delta^{13}C$ values for replicate extractions were less than 0.5 ‰, which indicates that the collagen extraction procedure was highly reproducible.

$\delta^{13}C$ values of collagen extracted from the bone samples are given in Table II. Samples up to and including 1000 AD range from -19.9 to -21.7 ‰ vs PDB, with an average of -20.9 ‰. All samples occurring after this time have significantly more enriched $\delta^{13}C$ values, indicating a shift in diet to include food with a higher ^{13}C content. Collagen $\delta^{13}C$ values after 1000 AD ranged from -10.4 to -15.8 ‰, with an average of -13.7 ‰.

Interpretation of Carbon Isotope Ratios

Since all known wild and domestic food plants in the eastern half of the United States prior to the introduction of corn used the C_3 pathway, the dramatic shift in collagen $\delta^{13}C$ values seen in these prehistoric bone samples could be interpreted only as the result of substantial maize consumption.

Controlled laboratory experiments with small mammals have shown that bone collagen is approximately 3 ‰ more enriched in ^{13}C than dietary carbon (15,18). In addition, by measuring collagen $\delta^{13}C$ values on prehistoric humans who could have ingested only carbon from C_3 plant sources, it has been inferred that human collagen is 5.1 ‰ more enriched in ^{13}C than the diet (19). Allowing for the +1 ‰ fractionation from NaOH treatment and the +5.1 ‰ fractionation between diet and collagen, the $\delta^{13}C$ value of the human diet prior to 1000 AD averaged -27.0 ‰, precisely the average value for C_3 plants. For the human bones from the period after 1000 AD, the same calculation indicates that the average $\delta^{13}C$ value for their diet would have been approximately -19.8 ‰.

When a diet consists of two isotopically distinct food sources for which the $\delta^{13}C$ values are known, it becomes possible to estimate the relative proportions of each by applying an isotopic mass balance equation:

$$D = C_3(x) + C_4(1-x) \qquad (2)$$

In this equation, D is the $\delta^{13}C$ value of the diet, C_3 is the $\delta^{13}C$ value of the C_3 plant material in the diet, C_4 is the $\delta^{13}C$ value of the C_4 portion of the diet, x is the proportion of C_3 material in the diet, and 1-x is the proportion of C_4 material in the diet. Assuming $\delta^{13}C$ values for C_3 and C_4 plants of -27.0 and -12.5 ‰, respectively, the human diet after 1000 AD in our study area averaged 50% C_3 material and 50% C_4 material or corn. For individual bones after 1000 AD, the estimates for corn consumption range from 35 to 72% of the diet.

Actual corn comsumption may be overestimated if the people in this sample consumed wild or domestic animals that also had consumed significant quantities of C_4 plants. ^{13}C enrichment from wildlife appears unlikely since the native vegetation consisted of C_3 plants. However, it is possible that wildlife could have fed on corn fields, and it also is possible that domestic animals could have been supplemented deliberately with corn. If either of these phenomena produced a ^{13}C enrichment in the animals, this enrichment then would have been passed along to the humans with the result of an overestimation of direct corn consumption. The magnitude of this effect presently is not known, but may be estimated in the future by analyzing the $\delta^{13}C$ values of collagen from associated animal bones.

Archaeological Implications

These results indicate that intensive maize agriculture in southeast Missouri and northeast Arkansas began around 1000 AD, that the shift to maize was rapid, and 35 to 72% of the diet may have consisted of corn after this time. While several studies report the presence of maize kernels at midwestern archaeological sites as early as 500 BC (38), our data provide isotopic evidence that maize could not have been a significant part of the human diet prior to 1000 AD. Our $\delta^{13}C$ values of bone collagen before 1000 AD are identical to measurements made on prehistoric humans from northern temperate zones where no C_4 plants could have been consumed (19). Data from nearby archaeological sites in Illinois show a similar pattern of $\delta^{13}C$ values through time, with significant maize consumption beginning between 1000 to 1200 AD (39).

Archaeologists generally have considered the Mississippian Period to be associated with intensive maize agriculture (40). Mississippian sites began to appear in our study area around 700 AD (41), so our data suggest that the Mississippian Period began several hundred years before maize agriculture became important. This interpretation is supported by our data from the Zebree site (Table II). Three of the samples from the earlier Big Lake Phase which dates to 900 AD show that corn was not a significant component of the diet at this time. By contrast, the Zebree sample from the Lawhorn Phase dates to 1200 AD and shows isotopic evidence for substantial maize in the diet (Table II). Thus, it appears that the hypothesized shift to maize agriculture in southeast Missouri and northeast Arkansas is not associated with the appearance of Mississippian culture, but occurs somewhat later in the chronological sequence.

Table II. Descriptions of human bone samples and results of carbon isotope analyses

Site	Burial number	Location	Temporal placement	$\delta^{13}C$
Scatters* (3RA19)	1	Randolph Co., AR	Late Archaic 3200 BC	-21.1
Lepold (23RI59)	--	Ripley Co., MO	Late Archaic 1980 BC	-21.7
Billy Moore (23BU26)	1	Butler Co., MO	Late Archaic 1000 BC	-20.5
McCarty (3PO467)	--	Poinsett Co., MO	Early Woodland 200 AD (Tchula Occupation)	-21.7
Christensen Cave	--	Pulaski Co., MO	Woodland 500 AD (Meramec Springs Complex)	-19.9
Nevins Cairn (23PU200)	2	Pulaski Co., MO	Woodland 500 AD (Meramec Springs Complex)	-20.1
Zebree (3MS20)	3F	Mississippi Co., AR	Mississippian 900 AD (Big Lake Phase)	-21.2
Zebree (3MS20)	6	Mississippi Co., AR	Mississippian 900 AD (Big Lake Phase)	-20.5
Zebree (3MS20)	3G	Mississippi Co., AR	Mississippian 900 AD (Big Lake Phase)	-21.2
Round Spring (23SH19)	1	Shannon Co., MO	Woodland 1000 AD (Meramec Springs Complex)	-20.7

Round Spring (23SH19)	2	Shannon Co., MO	Mississippian 1200 AD	-15.6
Lilbourn (23NM38)	16D	New Madrid Co., MO	Mississippian 1200 AD	-14.9
Zebree (3MS20)	4	Mississippi Co., AR	Mississippian 1200 AD (Lawhorn Phase)	-13.0
Turner (23BU21A)	21A	Butler Co., MO	Mississippian 1300 AD (Powers Phase)	-15.8
Turner (23BU21A)	28	Butler Co., MO	Mississippian 1300 AD (Powers Phase)	-13.2
Turner (23BU21A)	36B	Butler Co., MO	Mississippian 1300 AD (Powers Phase)	-14.1
Berry (23PM59)	22	Pemiscot Co., MO	Late Mississippian 1600 AD	-13.5
Campbell (23PM5)	52	Pemiscot Co., MO	Late Mississippian 1600 AD (Armorel Phase)	-10.4
Hazel (3PO6)	56	Poinsett Co., MO	Late Mississippian 1600 AD (Parker Phase)	-12.9
--	--	Butler Co., MO	19th Century Euro-American ca. 1880	-13.3

*The first number given is a code for the name of the state, the letters are an abbreviation for the county in which the site is located, and the second number identifies the archeological site within the county.

The Problem of Collagen Turnover

One of the unresolved questions remaining in the application of stable carbon isotope techniques to archaeological studies of human nutrition is the rate of turnover of bone collagen. It is well known that tissues and biochemicals in the human body are in a state of flux, i.e., being broken down and resynthesized constantly. Reassembly may occur with other breakdown products within the body, or with nutrients recently acquired through the diet. Thus, fluctuations in the isotopic composition of a diet will produce isotopic fluctuations in body tissues and biochemicals, dependent upon the turnover rates of the components in question. Some estimates of human bone collagen turnover are available using ^{14}C, introduced into the atmosphere by nuclear weapons testing, as a tracer (42,43). These studies indicate that bone collagen turns over very slowly, if at all. However, the long half-life of ^{14}C does not lend itself to the measurement of the rapid turnover that occurs in terms of hours or a few days.

As a continuation of our earlier studies on carbon turnover in animal tissue (17), we recently have measured the turnover rate of bone collagen in the gerbil using ^{13}C as a tracer. These data suggest that the half-life of bone collagen in the gerbil is approximately 60 days, and although this time is substantially longer than that of other gerbil tissues, it is much faster than would have been predicted based on turnover of human bone collagen using atomic-bomb generated atmospheric ^{14}C. Although animals with smaller body size, and therefore higher metabolic rates, would be expected to have much faster turnover rates than larger animals such as the human, these data suggest that bone collagen may be replaced faster than previously expected. This creates problems in the interpretation of human bone $\delta^{13}C$ values, i.e., what is the period of time over which collagen $\delta^{13}C$ values reflect the diet. This problem remains to be solved, but there seems little question that collagen $\delta^{13}C$ values do provide useful information concerning past diets.

Future Prospects for Stable Isotopes in Archaeology

It appears likely that the application of stable isotope techniques to the study of paleonutrition will continue. Carbon isotopes have been shown to be useful to trace the introductions of tropical C_4 crops into temperate regions where C_3 plants predominate. Other recent studies with carbon isotopes have demonstrated that seafoods are more enriched in ^{13}C than terrestrial foods, and that the proportions of marine vs land-derived foods can be estimated in prehistoric coastal people (44). Elements other than carbon may demonstrate isotopic distributions that will provide useful dietary information. For example, $^{15}N/^{14}N$ ratios of bone collagen may indicate the importance of legumes in the diet (30), and have been shown to be as useful as carbon isotopes in determining the relative importance of marine vs terrestrial food items (45). Because of the difficulties involved in determining diets of prehistoric humans from food remains at archeological sites, it seems likely that chemists and archaeologists will continue to collaborate in the development of new analytical techniques for the elucidation of dietary information from human bone.

Acknowledgments

This publication is supported in part by the USDA/ARS Children's Nutrition Research Center in the Department of Pediatrics at Baylor College of Medicine and Texas Children's Hospital and by a grant from the William and Flora Hewlett Foundation of Research Corporation. We would like to thank Drs. C. Chapman, J. Speth, D. Morse, and J. Rose for assistance in procuring human bone samples. We also are grateful to D. Nelson for performing the collagen extractions and the isotope analyses, Dr. C. Garza and M. Peter for the amino acid analyses, and E. R. Klein and M. Boyd for editing and preparing the manuscript.

Literature Cited

1. Klein, E.R.; Klein, P.D. Biomed. Mass Spectrom. 1978, 5, 321-330.
2. Klein, E.R.; Klein, P.D. Biomed. Mass Spectrom. 1979, 6, 515-545.
3. Bier, D.M. Nutr. Rev. 1982, 40, 129-134.
4. Klein, P.D. Fed. Proc. 1982, 41, 2698-2701.
5. Krueger, H.W.; Reesman, R.H. Mass Spectrom. Rev. 1982, 1, 205-236.
6. Boutton, T.W.; Smith, B.N.; Harrison, A.T. Oecologia 1980, 45, 299-306.
7. Boutton, T.W.; Arshad, M.A.; Tieszen, L.L. Oecologia 1983, 59, 1-6.
8. Craig, H. Geochim. Cosmochim. Acta 1953, 3, 53-92.
9. Hoefs, J. "Stable Isotope Geochemistry"; Springer-Verlag: New York, 1978; pp. 83-90.
10. Deines, P., in "Handbook of Environmental Isotope Geochemistry"; Fritz, P. and J. Fontes, eds.; Elsevier: Amsterdam, 1980. pp. 329-406.
11. Wong, W.W.; Benedict, C., Kohel, R. Plant Physiol. 1979, 63, 852-856.
12. Smith, B.N.; Epstein, S. Plant Physiol. 1971, 47, 380-384.
13. Whelan, T.; Sackett, W.; Benedict, C. Plant Physiol. 1973, 51, 1051-1054.
14. Kluge, M.; Ting, I. "Crassulacean Acid Metabolism; Analysis of an Ecological Adaptation"; Springer-Verlag: New York, 1978.
15. DeNiro, M.J.; Epstein, S. Geochim. Cosmochim. Acta 1978, 42, 495-506.
16. Teeri, J.; Schoeller, D. Oecologia 1979, 39, 197-200.
17. Tieszen, L.L.; Boutton, T.W.; Tesdahl, K.G.; Slade, N.A. Oecologia 1983, 57, 32-37.
18. Bender, M.M.; Baerreis, D.A.; Steventon, R.L. Am. Antiq. 1981, 46, 346-353.
19. van der Merwe, N.J. Am. Sci. 1982, 70, 596-606.
20. Price, T.D.; Kavanagh, M. Mid-Cont. J. Archaeol. 1982, 7, 61-79.
21. Zurer, P.S. Chem. Eng. News 1983, 61, 26-44.

22. Steyermark, J.A. "Vegetational History of the Ozark Forest"; Univ. Missouri: Columbia, MO, 1959; pp. 127-132.
23. Lewis, R.B. "Mississippian Exploitative Strategies: A Southeast Missouri Example"; Univ. Missouri: Columbia, MO, 1974; pp. 17-28.
24. Berger, R.; Horney, A.G.; Libby, W.F. Science 1964, 144, 999-1001.
25. Krueger, H., in "Proceedings of the Sixth International Conference on Radiocarbon and Tritium Dating"; U.S. Atomic Energy Commission: Oak Ridge, TN, 1965; p. 332.
26. Schoeninger, M.; De Niro, M.J. Nature 1982, 297, 577-578.
27. Wyckoff, R.W.G. "The Biochemistry of Animal Fossils"; Scientechnica, Ltd.: Bristol, 1972; p. 52-114.
28. Degens, E.T. in "Organic Geochemistry"; Eglinton, G. and M. Murphy, eds.; Springer-Verlag: New York, 1969; p. 314.
29. White, A.; Handler, P.; Smith, E. "Principles of Biochemistry"; McGraw-Hill: New York, 1968; p. 871.
30. DeNiro, M.J.; Epstein, S. Geochim. Cosmochim. Acta 1981, 45, 341-351.
31. Hakansson, S. Radiocarbon 1976, 18, 290-320.
32. Longin, R. Nature 1971, 230, 241-242.
33. Buchanan, D.; Corcoran, B. Anal. Chem. 1959, 31, 1635-1638.
34. Boutton, T.W.; Wong, W.W.; Hachey, D.L., Lee, L.S.; Cabrera, M.P.; Klein, P.D. Anal. Chem. 1982, 55, 1832-1833.
35. Des Marais, D.; Hayes, J. Anal. Chem. 1976, 48, 1651-1652.
36. Craig, H. Geochim. Cosmochim. Acta 1957, 12, 133-149.
37. Eastoe, J.E. Biochem. J. 1955, 61, 589-602.
38. Struever, S.; Vickery, K.D. Am. Anthropol. 1973, 75, 1197-1220.
39. van der Merwe, N.J., Vogel, J.C. Nature 1978, 276, 815-816.
40. Griffin, J.B. Science 1967, 156, 175-191.
41. Lynott, M.J. Southeast. Archaeol. 1982, 1, 8-21.
42. Libby, W.F.; Berger, R.; Mead, J.; Alexander, G.; Ross, J. Science 1964, 146, 1170-1172.
43. Stenhouse, M.J.; Baxter, M.S., in "Radiocarbon Dating"; Berger, R. and H. Suess, eds.; Univ. California Press: Berkeley, 1979; pp. 324-341.
44. Tauber, H. Nature 1981, 292, 332-333.
45. Schoeninger, M.J.; DeNiro, M.J.; Tauber, H. Science 1983, 220, 1381-1383.

RECEIVED January 30, 1984

14

Models for Carbon Isotope Fractionation Between Diet and Bone

HAROLD W. KRUEGER and CHARLES H. SULLIVAN

Geochron Laboratories Division, Krueger Enterprises, Inc., 24 Blackstone Street, Cambridge, MA 02139

Dietary evaluation using isotopic analyses of carbon in collagen from bone is an exciting new area of archaeological chemistry. Analyses of bone from herbivores, carnivores, and omnivores (including humans) suggest that a simple isotopic fractionation between dietary carbon and carbon in bone collagen may be an inadequate model for interpretation of results. Dietary carbohydrates are primarily metabolized for energy and their carbon is reflected mainly in the hydroxyapatite of bone. Dietary lipids are also important energy components of the diet. Dietary proteins, on the other hand, are utilized for protein (e.g. bone collagen) synthesis as needed and only excess amino acids are metabolized for energy. Herbivores, carnivores, and omnivores thus might have different isotopic fractionation models, each of which is presented. Biochemical evidence in support of these models is discussed.

Isotopic studies relating to nutrition and diet have originated from two diverse fields, bio-medical research and archaeology. Numerous studies have been reported by researchers in the areas of biochemistry and medicine, using either isotopically enriched compounds or the natural variations in isotopic abundances. Such studies usually involve a specific chemical as a tracer of biochemical pathways, and in these studies soft tissues or body fluids are analyzed. Recent work in nutrition has begun to examine the isotopic composition of carbon in macronutrients in diets and their disposition in body tissues (1,2). The isotopic composition of hard tissue (i.e., bone), however, has been largely ignored in biochemical studies.

Archaeologists are often interested in dietary considerations, but normally have only bone to represent the human organism, the soft tissues having decomposed. Dietary reconstruction using

0097-6156/84/0258-0205$06.00/0

artifacts or refuse is circumstantial. The interpretation of such data is tenuous because these preserved and recovered materials represent what was not actually eaten, they may not include evidence of what was eaten elsewhere, so it is difficult to determine the correct proportionality of food types using only the recovered materials. Bone, on the other hand, represents a stable tissue produced by the organism over a long period of time, and its composition should bear a definite relationship to dietary intake. Isotopic analyses of bone should reflect the variations in isotopic composition of the diet.

The primary focus of isotopic studies on human bone has revolved around the distinction between consumption of C_3 plant material and C_4 plant material. Some years ago, it was discovered that the C_3 (or Calvin) and the C_4 (or Hatch-Slack) photosynthetic pathways generated plant tissue with quite different ^{13}C abundances, an approximately 15 parts per thousand (o/oo) difference in the isotopic ratio (3). This isotopic difference between two types of plants is the main basis for most studies of human diets that have used stable isotopes of carbon as an analytical tool. Most plants in temperate areas are of the C_3 type, but corn (maize) is a C_4 plant and is of special interest to archaeologists because of the apparent dependence of many cultures on maize agriculture.

Serious efforts to understand the relationships between diet and the isotopic composition of bone began in the mid-1970s. Vogel and others (4), studied the distribution of C_3 and C_4 plants in southern Africa, and then the collagen in bones of African animals with known feeding patterns (5). Diet was readily correlated with isotopic signatures in bone collagen. At about the same time, DeNiro and Epstein (6) were performing controlled feeding studies on a variety of lower animals and on mice, to quantify the isotopic fractionation between diet and animal tissues. Both groups of investigators soon took up the challenge of human diets and began studying archaeological human bone from various sites of interest. Other research groups have joined the effort in the last few years.

Almost all of these investigators utilize collagen as the material analyzed, and make the assumption that the isotopic composition of collagen represents the total average dietary intake. It is generally espoused that there is a direct relationship between the ^{13}C of collagen and the ratio of C_3 to C_4 plants in the diet. Although it is recognized that marine foods and some other special diets might make the interpretation ambiguous with respect to the percentage of C_4 plants in the diet, the archaeological context of the bones can often clarify this point.

The progress of the past decade has been thoroughly reviewed by van der Merwe (7) and has led to a model where the ^{13}C of collagen in bone represents the isotopic composition of the plant food dietary base, adjusted for a fractionation of +5 parts per thousand between the diet and the collagen formed. Using this model, one can calculate the percentages of C_3 and C_4 plant foods in the diet and thereby draw important conclusions about food procurement patterns and agriculture development. An example of this conventional model, adapted from (7), is shown in Figure 1.

Atmospheric CO_2 is isotopically fractionated during photosynthesis to yield the characteristic carbon isotope compositions (see definition below) for C_3 and C_4 plants as shown in Figure 1. The analysis expected for a 50% mixture of the two types is also shown. Allowing for the observed shift of +5 o/oo between diet and collagen formed, the C_3 plant diet (-26.5 o/oo) would produce collagen with a value of -21.5 o/oo, while the C_4 plant diet (-11.5 o/oo) would produce collagen with a value of -6.5 o/oo. A 50% mixture (-19 o/oo) would produce collagen of -14 o/oo. This simple relationship appears to hold quite well for pure vegetarian diets, but might not be adequate to describe the isotopic effects in the more complex diets of carnivores and omnivores. We present here new models for all three dietary types, based upon analyses of both collagen and hydroxyapatite in bone (8), that take into consideration the actual macronutrients in the diet and their ultimate fates.

Fate of Dietary Macronutrients

Total diet usually consists of one or more actual foods, but the diet can be more accurately characterized by breaking it down into chemical groups termed macronutrients. The major macronutrient groups are carbohydrates, lipids, and proteins. These three groups comprise the bulk of any diet and will have the largest influence on the isotopic composition of carbon in tissues formed. Each macronutrient group has unique isotopic characteristics relating to its formation (9), and each group is utilized differently by the body.

Carbohydrates provide the largest single source of carbon in all but carnivorous diets. They are generally used promptly for energy metabolism or converted to glycogen for storage and later use. Other metabolic pathways are available if required, but most carbohydrates are ultimately converted to CO_2 which is transported largely as blood bicarbonate to the lungs and expired. Thus, tissues which incorporate carbon from blood bicarbonate will be influenced by the isotopic characteristics of carbohydrates used for energy metabolism. In bone, that tissue is hydroxyapatite, which incorporates carbonate ions during crystal growth. The carbonate ions are presumed to derive from blood bicarbonate (8).

Lipids are a second major source of carbon in most animal diets, perhaps the major source in some carnivorous diets. Lipids are metabolized for energy requirements, although somewhat more slowly than carbohydrates. Lipids tend to be depleted in ^{13}C by about 2 o/oo or more relative to carbohydrates in any particular food (data in 6,7,9). Virtually all lipid carbon is ultimately

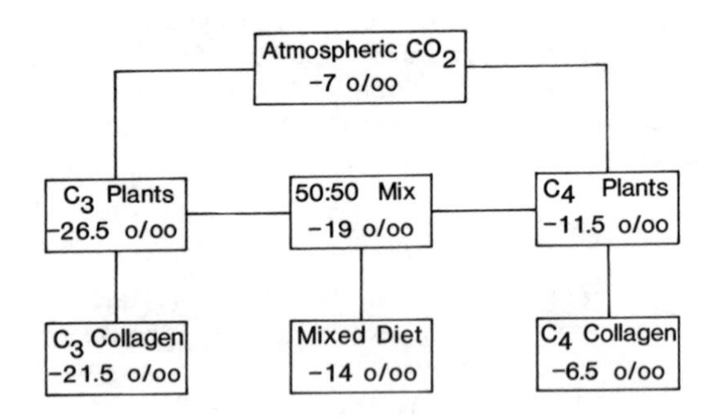

Figure 1. Model of ^{13}C in bone collagen produced from various plant diets (from data in 3-9).

metabolized for energy and is expired as CO_2 after passing to the lungs. In bone, therefore, the isotopic signal of lipids will appear primarily in hydroxyapatite.

Proteins are the third major carbon source in animal diets, and are essemtial for tissue growth. They are generally enriched by about 4 o/oo relative to carbohydrate in any particular food (see data in 6,7,9). Although animals cannot create totally new amino acids, pathways are available to modify existing amino acids, transamination of keto-acids being the most prevalent. While it is generally stated that most animals require 4 to 5% dry weight of protein in their diets to thrive (10,11,12), a simple calculation involving daily dietary mass, time of growth to adulthood, and protein content of the adult, reveals that less than 1% dry weight of protein in the bulk diet is actually required to satisfy net protein growth. The majority of protein in the diet of a healthy animal is used for the replacement of tissues being turned over and at least some of the protein originally in these replaced tissues is metabolized and excreted.

The requirements for net protein increase are usually met in the diets of healthy animals as evidenced by the growth of the animal. Although many animals suffer periods of starvation or of inadequate diet (10,11,12), during which their diets have less protein than required, these periods are times of tissue loss, not growth, and should have little or no effect on the isotopic composition of bone tissues. The amino acids required for bone growth may be derived directly from the diet if the appropriate amino acids are available; if not, they will be produced, as required, by transamination of keto-acids derived from the carbohydrate portion of the diet or from previously assimilated carbon. Essential amino acids must, of course, be acquired in the diet by humans.

Chemistry of Bone

Bone is composed of intertwined organic and inorganic components which lend this tissue its remarkable properties of combined strength and resiliency. Fresh air-dried bone is about 70% inorganic, 20% organic and 10% bound water.

The major organic component in bone is collagen, which constitutes some 90 to 96% of the dry fat-free organic matter. Collagen is a long fibrous protein, two-thirds of which is formed of only four amino acids. Collagen is about 33% glycine, 11% alanine, and 22% proline and hydroxyproline (13).

The inorganic matrix of bone is considered to be a calcium phosphate, more or less well crystallized as hydroxyapatite, which if pure, would have a formula of:

$$Ca_5(PO_4)_3(OH)$$

Actually, carbonate ion substitutions make up as much as 5% of bone apatite, mainly substituting for phosphate groups, although it is likely that a small amount exists as a substitute in the hydroxyl position or attached in labile surface positions on the crystals. These carbonate ions are presumably incorporated into the apatite structure from dissolved bicarbonate in body fluids during crystal growth.

The major sources of carbon for possible isotopic analysis in bone are collagen, which contains about 45% carbon, and hydroxyapatite, which contains a few percent of carbon present as carbonate ion substituted in the apatite crystals. These two carbon sources are derived from different pathways and different parts of the diet. Turnover of bone tissues is the slowest of all tissues in the body that have been studied, the mean residence time of carbon in collagen being at least 30 years (14).

Other carbon components of fresh bone have generally been decomposed and removed from archaeological bone, but extraneous carbon compounds may have been introduced. Interfering organic contaminants which are likely to occur in bones which have been buried for some time are roothairs, insect cuticles, fungal hyphae, and the humate decay products of a variety of other plant and animal compounds. Calcium carbonate carried into the sediment by rain or groundwater can percolate through a bone and deposit calcite which could interfere with the analysis of hydroxyapatite.

Fortunately, these types of contaminants can be fairly effectively removed by appropriate pretreatment techniques, leaving intact the apatite carbon and, if not decomposed, also the collagen. Our procedures have been derived from our experience in radiocarbon dating of bone, where complete removal of extraneous carbon is essential (15). The use of carbon derived from hydroxyapatite (8) for isotopic analysis has been disputed (16) on the basis that exchange of carbon from dissolved carbonate in groundwaters may have altered the isotopic composition of the hydroxyapatite. The data obtained from radiocarbon dating of bone, using carbon from hydroxyapatite, shows that such exchange is limited to a few per cent of the carbon content, if it occurs at all, and hence would not influence the isotopic composition of hydroxyapatite beyond normal analytical error (17). Analyses of hydroxyapatite from 3 million year old teeth give correct dietary interpretations (18). Another study, in progress in our laboratory, indicates that 30 million year old bones and teeth give isotopic analyses on hydroxyapatite that are interpretable in terms of a diet that was available and appropriate to the animals studied, and are different from the isotopic composition of carbonates in the enclosing rocks.

Analytical Procedures

Procedures for the preparation of bone samples for isotopic analyses are critical for obtaining reliable data. Archaeological bone for analysis is first thoroughly washed and wire-brushed to remove external material. If the sample is large enough, the bone is split and the contents of the marrow cavity are completely discarded. The fragments are then soaked overnight in Alconox, an alkaline lab detergent designed to remove bloodstains and other difficult organic contaminants. The soaking is repeated until no significant discoloration of the solution occurs overnight. The fragments are then rinsed and soaked overnight in 1N acetic acid to remove externally deposited carbonates. This is repeated until no observable reaction occurs.

The fragments are then washed free of acid, dried, and crushed to a smaller than 1 mm powder. The powder is again reacted with 1N acetic acid overnight to remove carbonates from within the bone matrix, longer if reaction is strong or continues. The treated bone powder is washed free of acid residues and dried. Pretreatment procedures have been tested by analyzing modern, uncontaminated bone both with and without the treatments. Equivalent results were obtained within analytical error. The pretreatments are essential for good results on contaminated archaeological bone samples.

About 1 gram of bone powder is demineralized, under vacuum, with 1N HCl. The hydroxyapatite dissolves, releasing CO_2 from its substitution positions in the apatite. The released CO_2 is recovered and purified by freezing and distillation through a trap held at -130°C which removes sulfur compounds and certain organic compounds which might also be released. This CO_2 is then analyzed as the apatite fraction.

Raw collagen, if preserved, remains insoluble and is washed by decantation. Distilled water is added, and a few drops of HCl to maintain acid conditions, and the collagen is dissolved by boiling. The hot broth is filtered through fiberglass filters, the dissolved collagen passing into the filtrate, while humic acids, rootlets, insect chitin, and other contaminants remain on the filter and are discarded. The filtrate is evaporated to a small volume, transferred to Teflon crucibles, and dried to crystalline gelatin in an oven at 70°C to prevent any decomposition and selective loss of carbon.

About 20 mg. of the dry gelatin are placed in a Pyrex tube with copper oxide. The tube is evacuated, sealed, and placed in an oven at 500°C overnight. The gelatin is combusted to CO_2, water and nitrogen gas.

The cooled tube is broken under vacuum, and the gases are separated for analysis. Water is frozen out in a dry ice bath, CO_2 is retained in liquid oxygen, and repurified by distillation for gelatin ^{13}C analysis.

Mass spectrometric analyses of CO_2 are performed on a Micromass 903 triple-collecting mass spectrometer, with digital data collection, as described elsewhere (19). Overall precision of individual analyses is better than +/- 0.2 o/oo.

Isotopic Notation

Because the variations in the actual isotopic ratio of carbon are very small, results are reported in "delta" notation, or parts per thousand variation in the $^{13}C/^{12}C$ ratio from that of the PDB standard. The PDB standard is a calcium carbonate marine shell and is quite rich in ^{13}C. Therefore isotopic analyses of most mammalian tissues are negative relative to PDB. Isotopic analyses are calculated according to the relationship:

$$\delta^{13}C = \left[\frac{(^{13}C/^{12}C)_{sample}}{(^{13}C/^{12}C)_{PDB}} - 1\right] \times 1000$$

Analyses are therefore reported in "per mil" units (o/oo), or relative difference of the $^{13}C/^{12}C$ in parts per thousand from the $^{13}C/^{12}C$ of the PDB standard.

Herbivore Diets

Herbivore diets consist only of plant materials and thus are the simplest diets to understand. Herbivores eat the more digestible portions of plants, so their diets consist of a preponderance of carbohydrates with only minor contributions of lipids and proteins from the plant material consumed.

Carbohydrates, including polysaccharides broken down to simpler sugars by bacterial action in the digestive tract, comprise the primary energy source for herbivores. In general, the simple sugars and starch in consumed vegetation, along with digestible sugars produced by bacterial decomposition of higher polysaccharides, satisfy most of the energy requirements of an herbivore. Since the carbon in these compounds is converted to CO_2, and ultimately respired, we would expect to see the isotopic signature of this part of the diet in the carbonate substitution in apatite of bone.

Lipids form a minor part of the carbon content of herbivore diets. As a result, despite the fact that they are used primarily for energy metabolism, their abundance is not sufficient to greatly influence the isotopic composition of blood bicarbonate. For this reason, the isotopic composition of lipids will not significantly affect the isotopic ratio of carbon in bone apatite.

Protein in herbivore diets is likely to be adequate in amount to satisfy the dietary requirements for net growth of an herbivore (approx. 1% of bulk diet), and perhaps also the requirement for tissue turnover. The distribution of individual amino acids, however, may not even be close to those required for proper tissue growth or the turnover of existing tissues. As a result, it would be expected that a substantial amount of amino acid synthesis would occur, and that the collagen of herbivores would closely reflect the isotopic composition of the herbivore′s diet.

A theoretical model for carbon isotope fractionation between herbivore diets and herbivore bone is shown in Figure 2A. This model shows the isotopic relationship of gelatin to that of hydroxyapatite, and the relationship of both to the original plant food diet. The gelatin seems to reflect the growth portion of the diet while the apatite seems to reflect the energy portion of the diet. The difference of +7 o/oo between apatite values and gelatin values is due in part to the fact that blood bicarbonate has a ^{13}C enrichment produced by the transfer of CO_2 from blood plasma to expired air. This effect has been experimentally verified and will be described elsewhere.

Herbivore diets usually contain adequate amounts of protein, as evidenced by the net growth of the animal, but the mixture of amino acids derived from vegetation is probably different from that required by the animal. Thus herbivores probably synthesize a large proportion of their required amino acids by transamination of keto-acids derived from the carbohydrate part of their diet. As a result, one would expect that the carbon isotopes of both apatite and gelatin in herbivores would show a direct relationship to the mixture of C_3 and C_4 plants consumed.

The actual results of isotopic analyses of herbivore bone from animals with a variety of diets are shown in Figure 2B. The line represents a direct relationship between isotopic composition of gelatin and that of apatite, with an offset of +7 o/oo due to the blood bicarbonate effect mentioned above. The results in Figure 2B are all from contemporary herbivores, and fit the model closely. Ancient herbivore bones give similar results (8). Thus we can use either carbon isotopes in gelatin, or carbon isotopes in apatite, to determine dietary intake in herbivores. For very old bones, collagen is usually decomposed (15), and only the apatite results can be obtained.

Carnivore Diets

Carnivore diets are very different from those of herbivores, consisting of meat and other animal tissues and almost no plant foods. The macronutient composition of such a diet would consist of a great deal of protein and lipid and almost no carbohydrate. The amino acid requirements for growth would probably be satisfied directly from the diet since the amino acid distribution in meat would be close to that required for growth. Little or no amino acid synthesis from carbohydrates or lipids would be required, therefore, and the carbon isotopes in carnivore collagen would be essentially the same as those in the protein of the prey animals eaten. Carnivore diets contain a large excess of protein beyond that required for growth and tissue turnover, and the excess is metabolized and used for energy.

Lipids provide the major source of energy in carnivores. They are not substantially involved in the synthesis of other biochemicals and are largely stored and utilized as needed for energy.

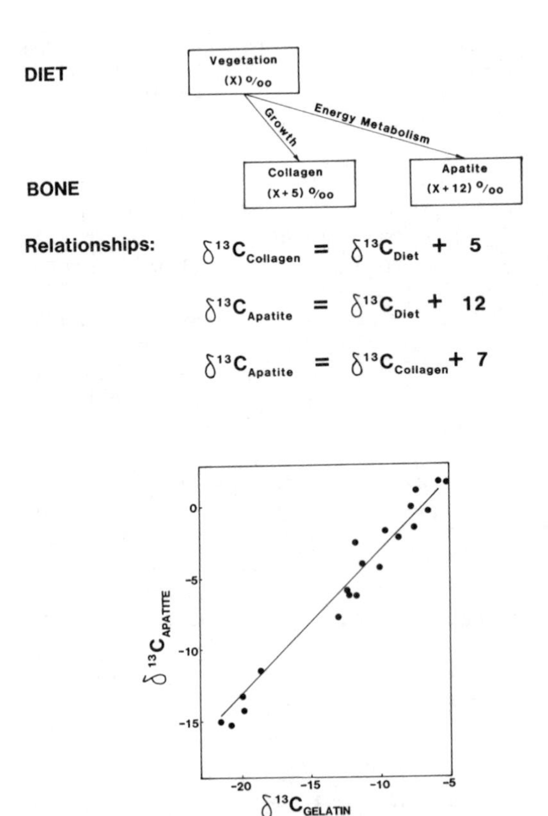

Figure 2. Top, theoretical model for carbon isotope fractionation between diet and bone in herbivores; and bottom, isotopic analyses of contemporary herbivore bones compared to the theoretical model (solid line).

Carbohydrates are almost negligible in pure carnivore diets, the only significant source being glycogen stored in the cells of the meat eaten. It is unlikely that this glycogen is sufficient in quantity to significantly affect the isotopic composition of blood bicarbonate or bone apatite.

A theoretical model of the isotopic composition of carnivore bone is presented in Figure 3A, embodying the principles just discussed. The isotopic composition of the meat eaten (usually herbivores) is controlled by the plants eaten by the prey. Gelatin in carnivore bone will directly reflect the protein or collagen of the prey (+5 o/oo relative to the diet of the prey) while the carbon in apatite will be largely controlled by the lipids in the prey, with only a lesser contribution from metabolism of excess protein. Since lipids in meat are known to be about -6 o/oo relative to protein in the same meat (5), and protein is +5 o/oo relative to the plant diet of the prey animal, the average ^{13}C of the energy portion of the diet would be about +1 o/oo relative to the plant diet of the prey. Allowing for the +7 o/oo fractionation associated with respiration (mentioned previously), we would thus expect a ^{13}C value of about +8 o/oo relative to the diet of the prey in the apatite of carnivores, rather than the +12 o/oo of herbivores.

The actual analyses of gelatin and apatite from a variety of carnivores are presented in Figure 3B. The line is derived from the theoretical model just discussed. The range of ^{13}C of gelatin is identical to that of herbivores, but the ^{13}C of apatite is shifted about 3 to 4 o/oo in a negative direction because of the high lipid consumption and its contribution to energy metabolism.

Omnivores (including Humans)

Omnivores, particularly humans, provide the real challenge at modelling the isotopic composition of bone as a function of diet. They have diets that are generally part plant and part animal, and neither part need be particularly selective of plant type or meat source. We have approached the understanding of isotope effects in human diets by examining the dietary macronutrients in both growth and energy contexts. Because of the complexity of the diet in omnivores, we will first examine the aspect of bone growth, which is reflected in the collagen matrix of bone.

The major portion of human diets, particularly ancient humans, is carbohydrate of plant origin. If all of the diet were of plant origin, the human would be an herbivore and fit the previously described model. In reality, human diets generally include a significant amount of meat or other protein-rich foods which contain approximately the assemblage of amino acids required for growth. If a human diet contains as little as 1% by weight of animal protein, the amino acid requirements for net bone collagen growth could be met directly by the diet and carbohydrates need

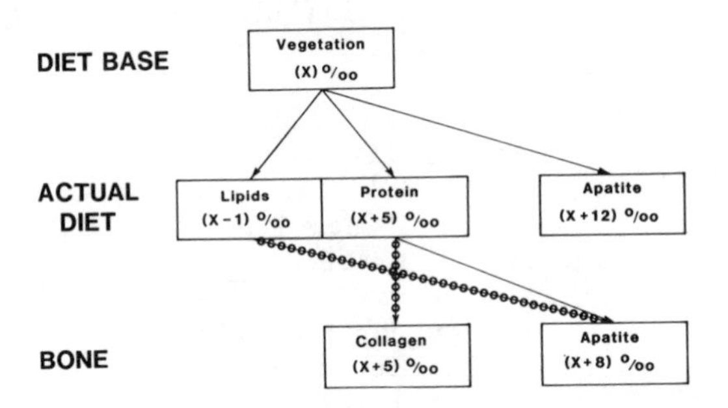

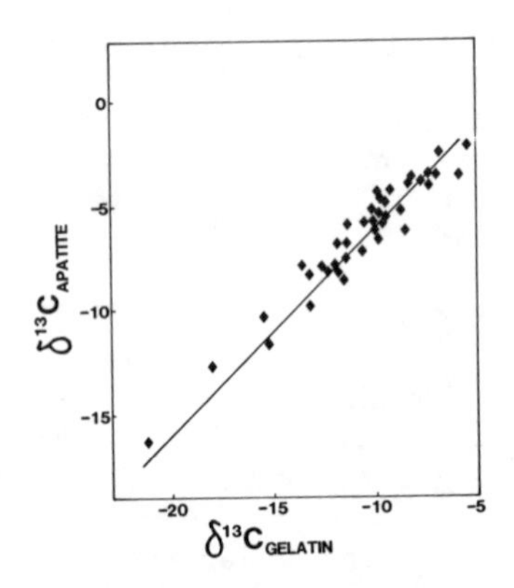

Figure 3. Top, theoretical model for carbon isotope fractionation between diet and bone in carnivores; and bottom, isotopic analyses of contemporary carnivore bones compared to the theoretical model (solid line).

not be significantly involved in collagen synthesis. If the diet contained 4-5% animal protein, even tissue turnover requirements would be fulfilled and there would be little or no need for involvement of carbohydrates in protein synthesis. In modern human diets, this has been demonstrated (2). The carbon isotopes of collagen thus may not reflect the diet in total, but only the meat portion of the diet. A theoretical model of isotopic relationships in collagen formation is shown in Figure 4A, indicating that the major factor controlling the isotopic composition of collagen is protein from the meat portion of the diet.

The energy metabolism model is quite different. In this instance, metabolism of carbohydrates provides the majority of the energy requirements. Lipids (from meat) provide a secondary energy source. Proteins from meat are a minor source of energy, only to the extent that they are not needed for growth. All of the macronutrients utilized for energy will be converted to CO_2 and transferred to the lungs for respiration. The isotopic characteristics of macronutrients used for energy will thus appear in the carbonate of hydroxyapatite. The isotopic relationships to be expected in human bone as a result of energy metabolism are shown in Figure 4B. The controlling part of the diet, for energy production, is plant-derived carbohydrates, with lesser contributions from lipids and only minor contributions from proteins.

If we examine the isotopic relationships to be expected in diverse human diets in light of the above models and the potential food sources, we can predict the isotopic compositions of human bone for each of these diets. Figure 5 shows the approximate isotopic compositions to be expected in human bone for eight different human diets which are reasonably common in modern man or in prehistoric societies. The ranges of isotopic composition shown are based on the typical ranges of isotopic composition of the macronutrients in each dietary type, and their probable proportions. The upper line is the herbivore line given in Figure 2B, and the lower line is the carnivore line given in Figure 3B. If our models are correct, the analyses of most human bones should fit the fields illustrated.

The actual isotopic analyses of over 200 human bone samples are shown in Figure 6, along with the dietary fields predicted in Figure 5. All of the samples are of archaeological origin, and most have associated evidence relating to probable diet. Over 98% of the samples fall within the dietary field expected on the basis of archaeological evidence and the remainder are close. Examples of seven of the eight hypothetical diets have been observed, the only exception being pure C_4 vegetarian diets.

Summary

Consideration of the macronutrients in various animal diets, their isotopic compositions, and the pathways by which they are utilized in the body, lead us to conclude that different models

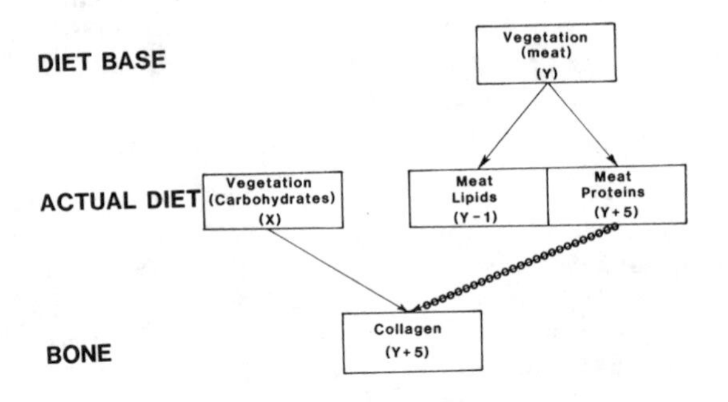

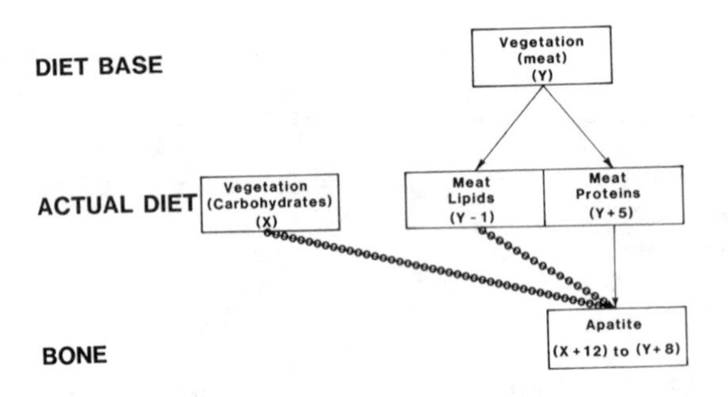

Figure 4. Theoretical models for carbon isotope fractionation between diet and bone in humans. Top, isotopic fractionation resulting from organism growth (major pathway emphasized); and bottom, isotopic fractionation resulting from energy metabolism (major pathways emphasized).

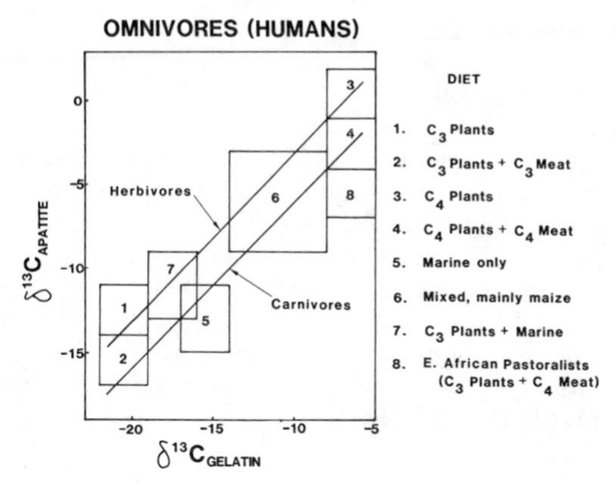

Figure 5. Expected ranges of carbon isotopic composition of human bone for eight possible diets.

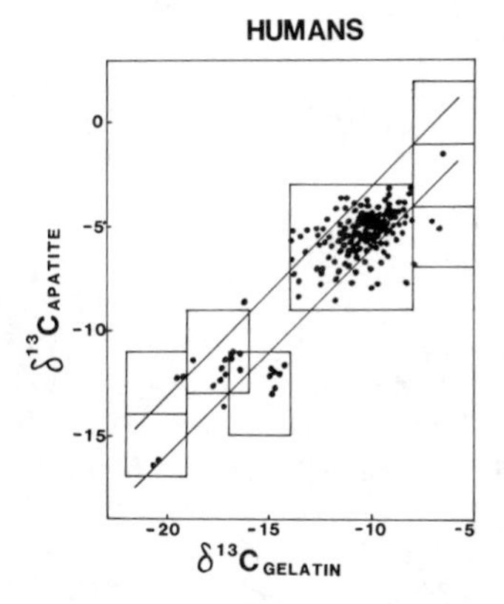

Figure 6. Carbon isotopic analyses of archaeological human bone compared to the theoretical model for herbivores, the theoretical model for carnivores, and the expected ranges of carbon isotopic composition of human bone for eight possible diets (<u>see</u> Figure 5 for identification of diet types).

for the isotopic composition of bone are required for herbivores, for carnivores, and for omnivores (including humans).

In herbivores, the plant food diet controls the isotopic composition of both bone collagen and hydroxyapatite, although isotopic fractionations exist in both phases.

In carnivores, bone collagen reflects the protein of prey animals directly, while the isotopic composition of apatite is largely controlled by lipids in the diet.

In humans, the meat protein of the diet usually controls the isotopic composition of bone collagen, while carbohydrates from plant foods usually control the isotopic composition of apatite.

Our models indicate that reconstruction of diets from isotopic analyses of bone cannot be accomplished solely by the use of the collagen fraction. The use of both collagen and hydroxyapatite phases of bone allows differentiation between energy and growth components of the diet, and therefore gives a considerable amount of detailed information about the total diet.

Literature Cited

1. Nakamura, K.; Schoeller, D.A.; Winkler, F.J.; Schmidt, H.-L. Biomed. Mass Spec. 1982, 9, 390-4.
2. Schoeller, D.A.; Koralewski, C.; Nakamura, K. XIth Int. Conf. Arch. Ethnol. Sci., Vancouver, Canada, August 1983. (In prep.).
3. Smith, B.N.; Epstein, S. Plant Physiol. 1971, 47, 380-4.
4. Vogel, J.C.; Fuls, A.; Ellis, R.P. South Afr. J. Sci. 1978, 75, 209-15.
5. Vogel, J.C. South Afr. J. Sci. 1978, 74, 298-301.
6. DeNiro, M.J.; Epstein, S. Geochim. et Cosmochim. Acta 1978, 42, 495-506.
7. van der Merwe, N.J. Amer. Sci. 1982, 70, 596-606.
8. Sullivan, C.H.; Krueger, H.W. Nature 1981, 292, 333-5.
9. Deines, P. in "Handbook of Environmental Isotope Geochemistry"; Fritz, P.; Fontes, J.Ch., Eds.; Elsevier: New York, 1980; Vol. 1, Chap. 9.
10. White, T.C.R. Oecologia 1978, 33, 71-86.
11. Mattson, W.J., Jr. Ann. Rev. Ecol. Syst. 1980, 11, 119-61.
12. Sinclair, A.R.E. J. Anim. Ecol. 1975, 44, 497-520.
13. Brown, C.H. "Structural Materials in Animals"; John Wiley & Sons: New York, 1975.
14. Stenhouse, M.J.; Baxter, M.S. Nature 1977, 267, 828-32.
15. Krueger, H.W. Proc. 6th Int. C-14 Conf., Pullman, WA, 1965, 332-7.
16. Schoeninger, M.J.; DeNiro, M.J. Nature 1982, 297, 577-8.
17. Sullivan, C.H.; Krueger, H.W. Nature, 1983, 301, 177-8.
18. Ericson, J.E.; Sullivan, C.H.; Boaz, N.T. Palaeogeogr. Palaeoclimatol. Palaeoecol. 1981, 36, 69-73.
19. Krueger, H.W.; Reesman, R.H. Mass Spec. Rev. 1982, 1, 205-36.

RECEIVED February 6, 1984

INDEXES

Author Index

Subject Index

R

S

T

Production by Paula Bérard
Indexing by Deborah Corson
Jacket design by Pamela Lewis

Elements typeset by Hot Type Ltd., Washington, D.C.
Printed and bound by Maple Press Co., York, Pa.

RECENT ACS BOOKS

"Bioregulators: Chemistry and Uses"
Edited by Robert L. Ory and Falk R. Rittig
ACS SYMPOSIUM SERIES 257; 286 pp.; ISBN 0-8412-0853-0

"Polymeric Materials and Artificial Organs"
Edited by Charles G. Gebelein
ACS SYMPOSIUM SERIES 256; 208 pp.; ISBN 0-8412-0854-9

"Pesticide Synthesis Through Rational Approaches"
Edited by Philip S. Magee, Gustave K. Kohn, and Julius J. Menn
ACS SYMPOSIUM SERIES 255; 351 pp.; ISBN 0-8412-0852-2

"Advances in Pesticide Formulation Technology"
Edited by Herbert B. Scher
ACS SYMPOSIUM SERIES 254; 264 pp.; ISBN 0-8412-0840-9

"Structure/Performance Relationships in Surfactants"
Edited by Milton J. Rosen
ACS SYMPOSIUM SERIES 253; 356 pp.; ISBN 0-8412-0839-5

"Chemistry and Characterization of Coal Macerals"
Edited by Randall E. Winans and John C. Crelling
ACS SYMPOSIUM SERIES 252; 192 pp.; ISBN 0-8412-0838-7

"Conformationally Directed Drug Design:
Peptides and Nucleic Acids as Templates or Targets"
Edited by Julius A. Vida and Maxwell Gordon
ACS SYMPOSIUM SERIES 251; 288 pp.; ISBN 0-8412-0836-0

"Ultrahigh Resolution Chromatography"
Edited by S. Ahuja
ACS SYMPOSIUM SERIES 250; 240 pp.; ISBN 0-8412-0835-2

"Chemistry of Combustion Processes"
Edited by Thompson M. Sloane
ACS SYMPOSIUM SERIES 249; 286 pp.; ISBN 0-8412-0834-4

"The Chemistry of Solid Wood"
Edited by Roger M. Rowell
ADVANCES IN CHEMISTRY SERIES 207; 616 pp.; ISBN 0-8412-0796-8

"Polymer Blends and Composites in Multiphase Systems"
Edited by C. D. Han
ADVANCES IN CHEMISTRY SERIES 206; 400 pp.; ISBN 0-8412-0783-6